enliten

The BTV/IP
Receiver Guide

May 2004

Edward Behan

Randy Palubiak

Disclaimer

While every precaution has been taken in the research and compilation of this Guide, the accuracy of the information contained herein cannot be guaranteed and Enliten Management Group, Inc. makes no representations or warranties, whether expressed or implied, as to the accuracies of the information contained herein.

This Guide may be amended or modified from time to time by Enliten Management Group, Inc., as deemed necessary without prior notification.

Copyright Information

Table of Contents

List of Figures

List of Tables

Executive Overview

Enliten Management Group

Enliten Management Group, Inc. (Enliten) is a consulting firm that helps enterprise organizations enhance and improve the delivery of their video and multimedia communications through the use of satellite-based delivery technologies. Enliten provides assistance and guidance in selecting satellite-based technology solutions and service providers, and in the implementation of new applications.

Many Enliten clients have implemented satellite-based networks for traditional business television (BTV) and interactive distance learning (IDL) applications and are looking to migrate to an Internet Protocol (IP) based platform in order to extend the reach of their networks. Others are considering IP as they build a business case for their initial satellite networks. We thank them for their inspiration in driving the development of this Guide.

In addition, we would like to express our appreciation to the service providers and manufacturers that participated in, and contributed to, the Guide. In particular, we thank those manufacturers whose products were not selected to be included.

Enterprise Users and Uses

Companies, organizations and associations continue to use the satellite delivery of information and programming on a full-time as well as ad-hoc basis. Many of the same applications that drove BTV in the early days are still key drivers today, including:

- **Time Management:** Maximizing executive and employee productivity by limiting the downtime associated with travel

- **Consistency of Message:** Delivering the same information simultaneously to a large, geographically-dispersed audience

- **Immediacy/Timeliness:** Distributing information in a timely manner

- **Increased Reach:** Delivering the message to a widely dispersed audience

- **Cost Savings:** Limiting travel costs by broadcasting to audiences in numerous locations

However, as the corporate use of video and multimedia increases, as well as key business drivers are expanding to include revenue-generating opportunities through advertising, marketing and promotional displays. In addition, video has gone beyond the traditional BTV and IDL television displays, as IP technologies enable news and information to be viewed on PCs in training rooms, at the desktop, via kiosks and at digital displays on the retail floor.

Today's IP technologies are capable of facilitating the delivery of rich media content to virtually anywhere at any time, efficiently and cost effectively. As the receiver is the central component of a satellite-based BTV/IP network, this Guide focuses its attention on the review of 17 models of receivers. This Guide is intended to provide enterprise communicators and stakeholders with information necessary and relevant to the decision-making process.

The following diagram illustrates various applications and display systems that IP technology can support. It is a non-technical representation, demonstrating that multiple programming channels and content can be distributed to numerous screens in various locations, including common areas, work areas or the retail floor. It is a tool to help identify where, when and how to roll out a new BTV/IP network, or enhance and improve an existing one to meet specific requirements.

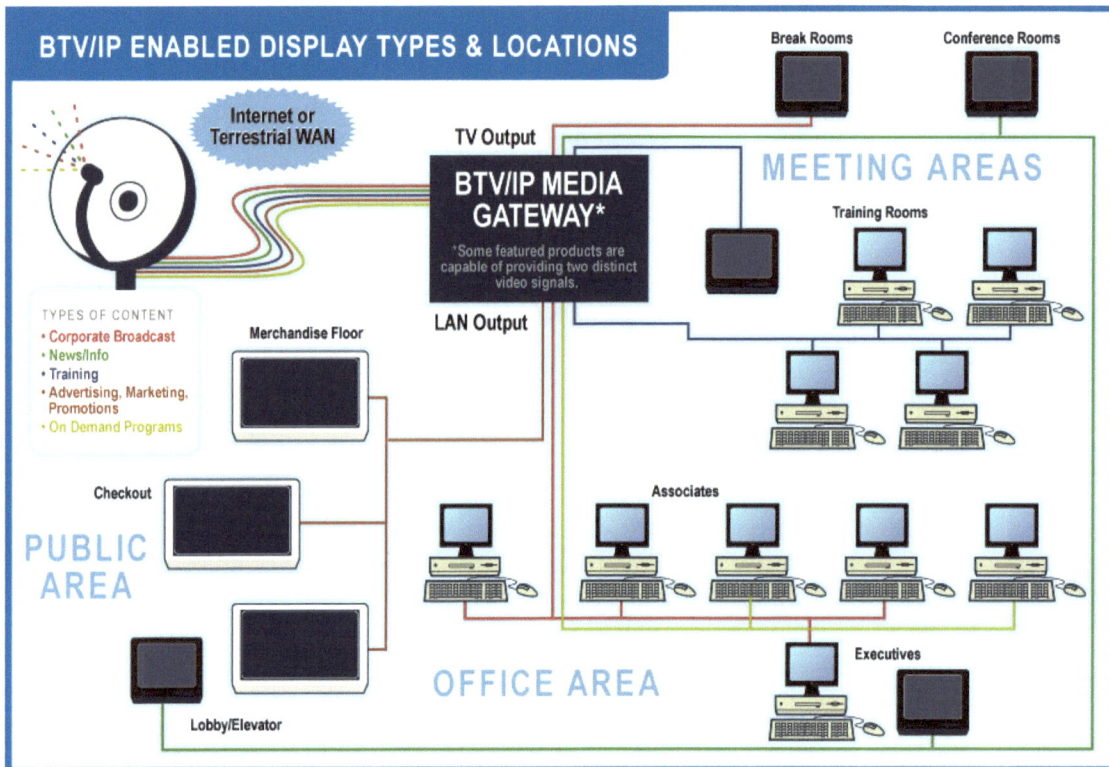

Figure #1:   BTV/IP Applications and Display Opportunities

Figures providing more detailed representations of BTV/IP technologies and key business applications are included in the *Network Architecture* section of this Guide.

It is not likely that any one enterprise will have the need or budget to justify the rollout of a BTV/IP network with all the display and playback capabilities shown in Figure #1, at least at the outset. When implementing a BTV network, history has shown that most enterprise users deploy applications in stages. Enliten and other industry specialists can provide expertise and guidance in identifying appropriate technology and developing a viable business plan to maximize the return on investment.

It is Enliten's belief that in the communications chain, network delivery is one of the least costly, if not the most cost effective, components. In addition, with the appropriate planning and guidance, the network can be one of the easiest of all elements of communications to implement and manage.

Where Does Enliten See the BTV Industry Heading?

Enliten believes that the time is right for BTV/IP solutions in the enterprise. A wide selection of satellite-based IP multicasting technologies and service solutions is available. The integrators and service providers have proven delivery systems and content management solutions, and have packaged these technologies into service offerings. Strategic relationships, partnerships and sales channels have been established between the IP receiver manufacturers and the service providers.

Furthermore, satellite service providers have relationships with training, e-learning and distance learning companies, as well as marketing and advertising companies, kiosk and display firms for digital signage, and point-of-sale/point-of-purchase (POS/POP) merchandising.

The traditional two-way VSAT providers, such as Hughes Network Systems, Gilat Spacenet and ViaSat are actively pursuing the BTV/IP solutions market. These providers employ IP in their transaction and data delivery networks, which is a natural migration to complement or support the one-way multicast of media-rich content.

In the past year, the number of enterprise users deploying BTV/IP solutions for video and multimedia applications has more than doubled. A few organizations are on their second technology platform. Still, the number of deployed networks is relatively small, with dozens across North America.

Other reasons for confidence in the growth of BTV/IP networks:

- Many existing BTV networks are operating with technology platforms that are nearing end-of-life and will soon be obsolete

- Computer networking technology – personal computers, in particular – found in the majority of enterprises are now robust enough to process useable video in addition to traditional data applications.

- Advancements in low data rate codec technology have improved the viewer's experience.

- Web-based video streaming has become increasingly common, driving the acceptance of IP video on local area networks by IT organizations.

- Satellite-based BTV/IP equipment, installation services and ongoing management and maintenance is more affordable than ever before, as is the space segment needed for the delivery

- With the improving economy, corporate budgets for video services are rebounding.

Additional Information and Guidance

The *Introduction* answers key questions and provides an overview of the content in the Guide.

Appendix A, Evolution of Business Television offers a history of the BTV industry, as well as information and a chronological storyline of industry service providers. *Appendix A* is intended to give enterprise stakeholders, especially those who are new to satellite-based delivery of video communications, a perspective on the industry, its longevity, integrity and reliability.

Appendix B, Glossary of Terms provides definition of terms used throughout this Guide.

1.0 Introduction

1.1 What is the Purpose of this Guide?

In recent decades, corporate and employee communications and training have become increasingly important to organizations and enterprises in building and maintaining healthy productivity, strong corporate cultures and continuous growth. Video-based technologies that support these efforts have grown increasingly sophisticated and ubiquitous.

In particular, satellite-based business television (BTV) and interactive distance learning (IDL) networks, employing one-way video transmission, have been widely adopted and successfully utilized by numerous organizations. However, many of these deployed technologies are approaching end of life and are on the verge of becoming obsolete. Simply put, they will soon need to be replaced.

A key question for both industry users and suppliers is, "Will next generation technologies be updated versions of old products, or will they offer new functionality?" Enliten strongly believes that the industry is poised to step into its next phase, based on the benefits of Internet Protocol (IP) technology.

Since the late 1990s, manufacturers and service providers across the industry have introduced efforts to deploy IP-based solutions. These initiatives did not gain traction due to a number of factors, most importantly, the absence of a single box solution to perform the principle service of BTV: delivery of live NTSC video to a television.

Other factors delaying the acceptance of IP-based solutions included: unfamiliarity with product manufacturers; poor video quality; prior investment in technologies that were stable; budget constraints; and technologies that required networked solutions be more sophisticated than were deployed in many enterprises at the time. This resulted in the impression that IP-based solutions consisted of unproven technologies that were not yet ready for prime time.

For all practical purposes, these issues and obstacles have been overcome. IP technologies today are past the bleeding-edge, leading edge phase and are moving into mainstream acceptance. Some of them have been deployed to thousands of sites. Sales and service channels are established. Essentially, BTV/IP has arrived.

This Guide specifically focuses on the key component of a BTV/IP network: the satellite receiver. It provides information and guidance on available, proven IP receiver technologies, the manufacturers who provide them, and touches on related components of BTV/IP network architecture.

1.2 What is BTV/IP?

BTV/IP stands for Business Television over Internet Protocol. Historically, IP has been the accepted protocol for the delivery of data over the Internet. Today, voice and video have also adopted IP, establishing IP as the universal protocol for the distribution of all information.

Unlike traditional BTV, where the digital video signal is distributed to televisions from the satellite receiver, BTV/IP video content can be distributed to both televisions and local area networks (LAN).

As such, BTV/IP can leverage computer networking technologies to enhance or extend traditional BTV services, with applications such as video on demand, video streaming to the desktop, dynamic digital signage, and real-time testing and measurement of viewer participation.

1.3   Why is IP Significant?

Most companies last considered or installed upgrades to their existing network infrastructure several years ago. Today, the capabilities of BTV/IP solutions are far beyond either the analog or digital video technology available at that time. Television and computer networking technologies continue to converge, and Internet Protocol is the foundation on which the convergence is based.

As local area networks grow more robust, computer processing more powerful and video compression quality higher at lower data rates, enterprise-wide video usage will become increasingly more prevalent.

Once users understand the new BTV/IP model they will recognize how to leverage its benefits throughout their organization.

1.4   Who is this Guide Written For?

This Guide is written for any organization that has an existing satellite-based business television network, as well as those considering the deployment of such a network. It is intended for any stakeholder in the organization involved in this decision-making process.

The purpose of this Guide is to help assure end-users that viable solutions to their organizations' communication delivery challenges exist and to introduce them to the landscape of available options. It is a starting point to help decision-makers save time and money as they research their next generation video delivery system.

1.5   What are the Benefits of BTV/IP to an Enterprise?

BTV/IP provides numerous benefits over traditional terrestrial and satellite-based business television and interactive distance learning networks, including:

- Extends the reach for corporate communications and training through services such as video on demand and IP video streaming
- Provides real-time control over corporate messaging
- Simplifies employee scheduling issues
- Allows for self-paced learning
- Improves measurement of viewer participation
- Facilitates and supports digital signage and other retail applications

- Provides associates with access to archived video content
- Reduces or eliminates the need to rebroadcast programming, improving the utilization of space segment
- Enables large file transfers of text and information
- Provides centralized control over video assets

1.6   Why Does the Guide Focus on the Receiver?

More than any other component, the receiver often represents the largest capital expenditure of a typical satellite network. The receiver's reliability and performance has had a major impact on the operational success of BTV networks. This is truer with BTV/IP, where the receiver must seamlessly integrate with corporate local area networks.

1.7   What Products are Included?

Enliten conducted extensive research of the various IP products available globally to identify which receivers meet the requirements of the satellite-based BTV/IP enterprise market.

The BTV/IP receivers included in this Guide are listed below in Table #1. Enliten has deemed these products relevant to the applications and needs of the enterprise user. They are segmented into the following categories:

- BTV/Media Gateway products designed exclusively for this market, offering:
    - Playback of NTSC video to a television
    - Streaming of IP video to a LAN
    - Hard drive for storage
    - Robust router functionality
    - Middleware for BTV-specific applications

- IP Satellite Routers
    - Satellite receiver with robust router functionality, intended for deployment in a large enterprise (to interface within a LAN) environment

- IP Satellite Receivers
    - Satellite receiver with an IP network interface, intended for the Small Office Home Office (SOHO) environment

Most of these BTV/IP receiver products are currently deployed in the field. They are represented through established sales channels and can be purchased directly from the manufacturer by self-managed networks.

ONE-WAY SATELLITE RECEIVERS INCLUDED IN THIS GUIDE

Adtec	Edje-L
Helius	2500-S, 1500S
International Datacasting	SFX2100, SRA2100, SRA2000
Ipricot	S1100, S1000, SC+, S500+
Mainstream	Data DVB+
Novra	S75, SSP 100
Skystream Networks	EVR-7000, EMR-5500, EMR-1600
Wegener Communications	iPump

Table #1 – One-Way Satellite Receivers Included in this Guide

Some of the products in this Guide do not meet the requirements of any one category. These products have been placed in the category deemed most appropriate.

Some IP receivers on the market do not meet the criteria for inclusion in this report, including:
- Those not available as a stand-alone box, but tied to a private label service
- Those not known to be deployed in the U.S. or Canadian market
- Two-way products, which typically are chosen based on factors not related to BTV/IP applications
- Video satellite receivers - traditional BTV integrated receiver/decoder (IRD) - with IP data ports that do not have router functionality

We anticipate that other manufacturers and their receivers will enter the BTV/IP space, just as some of those included in this guide are likely to depart due to market consolidation, changes in business strategy, and market acceptance.

1.8   What is the Methodology for Product Comparisons?

Enliten conducted interviews with the product manufacturers regarding their products' functionality and capabilities. Many service providers and end-users familiar with the selected products were also interviewed.

Next, Enliten compared commercially available specifications of the selected products and researched product details not found in the specifications.

Finally, the respective manufacturers reviewed the findings for accuracy only; their responses to the findings had no bearing on the comparisons presented in this Guide.

1.9   What are the Other Components of a BTV/IP Network?

In developing services for a BTV/IP network, additional components besides the satellite receiver need to be considered. These components are addressed in the *Network Architecture* section and include IP encapsulators; low data rate encoders; network management systems; remote hardware;

peripheral storage devices, dedicated players (for digital signage applications); interactive distance learning systems; and return path interfaces.

1.10   What is Enliten's Outlook for BTV/IP Services?

Enliten believes that the deployment of enterprise-wide video communications and applications will continue to increase, and that BTV/IP satellite-based services will play a pivotal role in driving this growth. It is inevitable that BTV/IP programming will be distributed through an organization's LAN and ultimately reach the desktop.

Enterprises who properly plan and implement BTV/IP services will succeed in improving the quality and effectiveness of communications and training throughout their organization, maximizing productivity and profitability.

2.0 Benefits of BTV/IP Technologies and Services

The principle benefit of BTV/IP services is that they allow the enterprise to maintain employee productivity while delivering corporate communications and training programming with greater flexibility. Compared with traditional satellite-based BTV networks, BTV/IP service reduces the need for centralized coordination of programs and events, the booking of meeting rooms and the scheduling of employees to participate. With BTV/IP services such as video-on-demand (VOD) and streaming video, all training and communication programming can be delivered directly to each employee's desktop.

In addition, IP technology allows the satellite network to offer new services to the enterprise. One of the most promising areas is in retail communications, where satellite-based digital signage services offer retailers real time control over messaging, allowing them to improve the customer experience and generate or increase revenues.

Another major benefit of BTV/IP technology is that the satellite network can act as an overlay to the corporate terrestrial Wide Area Network (WAN). These overlay networks allow the satellite link to be used for applications such as large file transfer and transport of other non-mission critical data. This reduces network congestion and improves the quality of service for terrestrial-based applications that are often mission critical, such as inventory management and credit card verifications.

The overlay approach also provides employees working in a SOHO environment the ability to enjoy secure, high-speed access to the corporate network, while reducing the requirements and costs of terrestrial connectivity from these locations.

2.1 A Proven and Integrated Technology

Although the number of networks using BTV/IP technology & services in the U.S. remains relatively small, the technology is becoming increasingly prevalent. It has proven to be reliable and has been integrated effectively into both enterprise LANs and WANs across North America.

The manufacturers of IP satellite receivers, although not as familiar to many in the BTV industry as Scientific-Atlanta and Motorola, have been producing large quantities of satellite receiver products for years. For example, Wegener Communications has been the sole manufacturer of satellite receivers for MUZAK's in-store audio, for over 20 years. Technology from Helius, Inc. has been found in VSAT solutions since the mid-1990s.

Early attempts at BTV/IP solutions were not successful due to technical issues that made integration of these solutions difficult. One of the major advances in resolving these issues has been the adoption by virtually all manufacturers of the Linux operating system. This was not the case with many of the early entries into BTV/IP that attempted to run on various Windows platforms that proved unreliable. These early problems forced some manufacturers to run on Unix-type operating systems. Although reliable, the cost of these Unix products was greater than the market was willing to support. Linux offers manufacturers an operating system that is both reliable and affordable.

BENEFITS OF BTV/IP TECHNOLOGY

BTV/IP Technology expands the reach and accessibility of Corporate Communications and Training Networks.

ISSUE	TRADITIONAL BTV CAPABILITIES	BTV CAPABILITIES
Employee Scheduling	Live Via Satellite	Live Via Satellite Video On Demand
Viewing Display	Television	Television Computer Monitor
Viewing Environment	Meeting/Training Room	Meeting/Training Room Desktop Retail Floor
Interactivity	Telephone Telephone Bridge Keypad Response System	Telephone Telephone Bridge Keypad Response System WAN/Internet
Collateral Materials	Hard Copy	Hard Copy Electronic Copy

Table #2 – Benefits of BTV/IP Technology

Sufficient BTV/IP networks have been deployed to allow the majority of integration issues to be resolved. This has been critical in helping BTV/IP services meet with approval of two key groups: IT departments and content viewers. IP satellite receivers meet the functional requirements of most corporate LANs. Bandwidth issues related to the transport of video over a LAN remain a concern of IT departments. However, the functionality of the BTV/IP technology is typically not an issue.

The user interface has been optimized by the integration of standard media players and middleware, such as Electronic Program Guides (EPG), requiring users to simply "point and click" to view programming, thereby rendering BTV/IP technology transparent to the end user.

2.2   Corporate Communication and Training Networks

The BTV/IP Media Gateway class of satellite receivers offers several significant improvements over traditional BTV satellite receivers. These technological improvements overcome the constraints of time and location to provide viewers with greater flexibility over when and where they view content.

The addition of a hard drive to the satellite receiver allows video content to be stored on the receiver at any or all remote network locations. This permits on-demand viewing on either a television or a networked computer. The benefits of Video On Demand service address many scheduling issues as discussed in Section 2.3.

Ethernet ports, router functionality, and extensive middleware associated with all of the products contained in this report enable the satellite receiver to become the point at which video fully integrates onto corporate LANs. This allows video to be streamed to the desktop, where many employees spend the majority of their time and where corporate information is regularly transferred. Further benefits of *Streaming to the Desktop* are detailed in Section 2.4.

Network Management Systems for the administration of video assets have also greatly improved over the past three years. All functionality found in traditional BTV technology platforms such as Scientific Atlanta's PowerVu Command Center can be found in the systems offered by some BTV/IP manufacturers. In addition, the functionality of IP Multicast software, such as Kencast, has been fully integrated into these next generation management systems. Lastly, with the use of a terrestrial return path, these systems can monitor the status and content of remote sites and report many abnormalities to the hub before remote sites identify them.

Section 2.5 discusses *Retail Communications Networks* where some manufacturers offer additional upgrades to these network management systems, allowing for the creation of play lists, reporting of playback data, and ad insertion.

Section 2.6 discusses satellite as an overlay to terrestrial wide area networks. These networks allow large file transfers to deliver support documentation, training manuals, pricing and retail/product information to the viewing location over satellite, reducing shipping costs for printed materials and improving QOS on the terrestrial network.

2.3   Benefits of Video-On-Demand for Corporate Communications and Training Networks

The combination of VOD and streaming video to the desktop technologies allows BTV/IP networks to expand their reach and provide enhanced benefits for corporate communications and training applications.

Video-on-demand essentially solves the issues associated with viewers watching a particular program. A partial list of benefits of VOD systems include:

- Reduced employee scheduling issues
- Reduced rebroadcasting of programs, resulting in either lower transmission costs or additional capacity for other programming
- Reduced requirements for peripheral recording devices such as videocassette recorders

- Viewer access to archived materials
- Real time control of archived materials
- Space segment reductions allow for increased targeted or niche programming
- Self-paced learning
- Increased viewership
- Improved measurement of viewer participation

Scheduling of employees to view "live" satellite broadcasts is often problematic. Many corporations with BTV networks are located across several time zones and have employees working various shifts. As a result, many employees are not scheduled to work at the time of broadcast. Staff performing mission critical operations and other assigned functions, as well as those who are traveling or absent from work, further limit the reach of traditional BTV networks.

It should be noted that the expense associated with the loss of employee productivity during scheduled corporate broadcasts – not to mention other related expenses – could be significant.

In a traditional BTV service, programming is rebroadcast over satellite to reach some of the viewers unavailable for the original broadcast. These rebroadcasts tend to have limited effectiveness since not all viewers are available at the scheduled rebroadcast times. In addition, for those who do not have space segment on a full-time (24x7) basis, this is often a costly proposition. VOD overcomes these issues by storing the program content on the satellite receiver, reducing space segment costs and allowing viewers to watch the programming at their convenience, thereby increasing audience participation.

Previously, videocassette recorders (VCRs) had been a valuable tool for BTV networks in overcoming scheduling issues. With IP-based technology, VOD provides a significant improvement, allowing programs to be stored on the LAN. Whereas VCRs presented administrative challenges, the on-demand approach eliminates the high degree of human involvement and operator error. There is no need to load, unload, label, and archive the tapes; nor any distribution and disposal issues. VCRs can continue to serve a complementary, supplemental purpose in BTV/IP networks, without the need to rely on videotaping material.

A major benefit of VOD services is related to archived materials and the administration of content. BTV/IP technology allows programming to be archived on servers at each remote site. The amount of materials that can be archived at each site is scalable to meet any network requirements. Access to the archives can be grouped and tiered to secure proprietary content. Reports on which viewers have requested which programming can be obtained. Programming which has reached its end of useful life can be remotely deleted from the archives.

BTV/IP services are designed for VOD archives to be centrally administered. This allows for greater control of corporate video assets, helps protect content from reproduction and redistribution, allows for obsolete materials to be disposed of electronically, and reduces the costs associated with administering these assets.

In addition, VOD enables organizations to expand the types of programming available to viewers. This is because VOD programming does not have to be transmitted "live" over the network. Instead, a particular program could be pushed to the storage device at the remote location as data, using only a fraction of the bandwidth necessary for a "live" transmission. The number and/or amount of unique programs is limited only by storage capacity, which is infinitely scalable.

Niche programming targeted to a subset of an organization can now be cost justified due to the greatly reduced space segment charges associated with the transmission of VOD programming. Coupled with this, is the reduction of expenses associated with scheduling employees to watch a broadcast at a specific time and the ability to have the content viewed repeatedly without additional transport charges or network congestion. The net result is a lower cost per employee for corporate communications.

Self-paced learning is another important benefit of VOD services. In both the corporate communications and training environment, content is regularly presented at a pace not well suited for some viewers. VOD services let the viewers control their own pace, with the ability to pause, move forward or back within a program. This results in improved learning and more efficient use of time.

Video-on-demand services can track and report viewer participation. This improved measurement of viewer participation is far more reliable than traditional BTV that often involve sign-in sheets at each remote location that must then be manually logged and reported to a central location. This is also an important factor in measuring which programming is most sought after by viewers, allowing for more targeted and effective communications.

The network architecture of how VOD services are implemented is contained in Sections 3.8 *Video-On-Demand to a Television* and Section 3.9 *Video-On-Demand to the Desktop.*

2.4   Benefits of Streaming Video to the Desktop for Corporate Communications and Training Networks

As mentioned in the previous section, the combination of VOD and streaming video to the desktop technologies integrated with BTV allows BTV/IP networks to expand their reach and provide the greatest benefits of all BTV/IP services for corporate communications and Training.

Whereas VOD addresses when a viewer watches a particular program, streaming video to the desktop addresses where a viewer watches the programming.

Benefits of streaming video to the desktop include:

- Increased participation during live broadcasts
- Reduced scheduling requirements for viewing environments
- Enhanced interactivity via WAN, Internet, or telephone
- Reduced cost by allowing presentation materials to be viewed electronically
- Improved learning over web-based e-learning, due to more dynamic content over higher bandwidth connections

Traditional BTV networks typically display programming on televisions or video monitors located in large open areas such as conference rooms or employee cafeterias. Viewing in this environment will remain valuable to enterprises using BTV/IP technology, and all of the Media Gateway class BTV/IP satellite receivers contained in this guide supports these applications.

Streaming of corporate communications to the desktop increases the audience for live broadcasts by allowing participation of those employees who are required to remain at their workstations. This increases the audience for corporate broadcasts, reduces the need to schedule large viewing environments and the need for rebroadcasts, and allows for better business continuity during these events.

Another major benefit of streaming video to the desktop is interactivity. Viewing at the workstation allows viewers to interact with the programming by using either their computer or telephone. The tools for both are very familiar to most viewers and require no learning curve. Quality issues often associated with return audio from larger remote meeting spaces are greatly diminished. In addition, some viewers may be more likely to ask a question using email or telephone than by standing in front of fellow associates and speaking into a microphone.

For both corporate communications and training applications, conducting the session at the desktop allows presentation materials to be viewed electronically. This results in obvious savings to both printing and shipping costs.

Due to the dramatic bandwidth differences of satellite transmission compared to web-based transmission, satellite-based BTV/IP services provide improved learning over web-based e-learning or web-cast corporate broadcasts by providing the viewer with more dynamic, higher resolution video content.

Section 3.7 *Live Video over Satellite to the Desktop* discusses the network architecture for these services.

2.5   Benefits of BTV/IP Technology for Retail Networks

BTV/IP technology has been effectively deployed in retail networks. These networks provide services that include digital signage, Advertising, Merchandising, & Promotion (AMP) displays, point-of-purchase displays, and even in-store audio systems. It should also be noted that these applications are not limited to retail environments. Many organizations utilize digital signage successfully as a form of internal communications.

Digital signage and other retail communication applications are typically intended to inform, influence, and entertain viewers. When successfully implemented, these services create sustainable and measurable value relative to branding, incremental sales, experiential retailing, private label merchandising, and trade dollar leverage.

Traditionally, content for retail networks has been distributed using hard media such as VHS tape and DVD. This method of transport has significant limitations. Implementing BTV/IP technology as the solution overcomes these limitations.

The benefits of a satellite-based transport solution for retail communications include:

- Immediacy
- Centralized control
- More dynamic media
- Networked solution
- Enhanced viewer experience
- Enhanced playback reporting

Perhaps the greatest benefit of the integration of BTV/IP technology with digital signage is immediacy. Compared with the traditional method of hard media content, satellite allows all of the displays in a network to be updated simultaneously within seconds. This provides retailers and other organizations the ability to keep messaging current, which improves the effectiveness of communications and the viewer experience, and helps drive additional revenues.

BTV/IP technology also provides networks with centralized control. This control enhances flexibility in programming selection and playback rotation, allowing network optimization for maximum return on investment.

When dedicated media player technology is combined with BTV/IP technology, the results can be dramatic. Layered - as opposed to pre-rendered – content allows graphical information to be incorporated with the video content to produce more dynamic media. This graphical information requires considerably less network bandwidth compared with pre-rendered content. Video content can be stored in the satellite receiver and streamed to the display, while graphical information, such as stock tickers, can be streamed live over the satellite using a small fraction of the bandwidth.

There are many other benefits associated with the networked solution enabled by BTV/IP technology. Many retailers find the most important of these to be enhanced playback reporting. This benefits retailers by helping assure that playback statistics are reliable, allowing for advertiser trade dollars to be fully leveraged. In addition, this feedback – coupled with sales statistics – provides the necessary data to continuously measure and improve the effectiveness of the retail network.

2.6   Benefits of Wide Area Network Overlay Services

Lastly, since a satellite-based BTV/IP network uses the same internet protocols as terrestrial WANs, the BTV/IP network can reduce traffic by acting as an overlay to the existing enterprise WAN. Any broadcast or multicast application can be routed over the BTV/IP network to optimize performance and improve the quality of service for mission critical applications.

Enterprises with a large SOHO base can benefit from BTV/IP technology's ability to deliver high speed Internet access to remote locations. The benefit is a reduction in the terrestrial WAN connectivity infrastructure requirements, lower potential cost, and improved network security.

In addition, enterprises can implement satellite-based BTV/IP technology as an overlay to a terrestrial WAN to provide high speed Internet access and other IP applications, to reduce terrestrial WAN requirements, transfer large files and relieve network congestion.

2.7 Summary

The following table summarizes some of the benefits of BTV/IP services and technology in the enterprise:

BTV/IP TECHNOLOGY > SUMMARY OF BENEFITS	
Video On Demand	• Reduced employee scheduling issues • Reduced rebroadcasting of programs, resulting in either lower transmission costs or additional capacity for other programming • Reduced requirements for peripheral recording devices such as videocassette recorders • Viewer access to archived materials • Real time control of archived materials • Space segment reductions allow for increased targeted or niche programming • Self-paced learning • Increased viewership • Improved measurement of viewer participation
Streaming Video to Desktop	• Increased participation during live broadcasts • Reduced scheduling requirements for viewing environments • Enhanced interactivity via WAN, Internet, or telephone • Reduced cost by allowing presentation materials to be viewed electronically • Improved learning over web-based e-learning, due to more dynamic content over higher bandwidth connections
Retail Communications	• Immediacy • Centralized control • More dynamic media • Networked solution • Enhanced viewer experience • Enhanced playback reporting
WAN Overlay Services	• Improved WAN quality of service • High speed internet access to SOHO environment • Reliable large file transfer at lower costs

Table #3 – BTV/IP Technology – Summary of Benefits

3.0 BTV/IP Network Architecture

The Receiver is one key component that distinguishes conventional BTV from BTV/IP technologies. Conventional BTV satellite receivers decode live broadcast transmissions for display on a television. The BTV/IP satellite receiver streams the video content through a local area network (LAN) for delivery to the desktop and a computer monitor. Some have the ability to playback NTSC video to a television, as well.

This Guide classifies BTV/IP satellite receivers in three categories, deemed relevant to the applications and needs of the enterprise:

- The BTV/IP Media Gateway
- The IP Satellite Router
- The IP Satellite Receiver

The receivers in each of these categories are designed for specific and distinct applications.

The BTV/IP media gateway offers the widest range of BTV/IP services. These satellite receivers have the ability to both decode video for television display *and* stream video over LANs. The hard drives built into these units provide VOD services. These units also have sophisticated router functionality for integration with large LANs and advanced middleware, providing viewers with user-friendly tools for accessing video content.

IP satellite routers and receivers offer a more limited range of BTV/IP services. In particular, they lack hard drives for storage and TV decoders for television viewing, and are limited in BTV/IP applications middleware. The difference between IP satellite routers and IP satellite receivers is that the routers can provide greater data throughput and can be seamlessly integrated onto a large enterprise LAN, while receivers have less data throughput and are designed for integration in typical SOHO environments.

Any of these products – or a combination of them – may be appropriate for a given enterprise satellite network, depending on the specific services provided by the organization. Today, enterprises have successfully deployed networks using products from all three of these categories.

BTV/IP technologies offer a wider range of applications than their predecessor, the standard BTV integrated receiver/decoder (IRD), and therefore the network architectures can be more complex. BTV/IP networks require larger, more sophisticated head ends that include products such as an IP encapsulator (IPE) and more robust network management systems. They also need to function reliably as part of the LAN at each remote site in the network and provide viewers with the necessary tools to display different forms of video content on a variety of displays.

This section is intended to provide readers with a general understanding of the network architecture and functionality of BTV/IP technologies.

3.1 Traditional BTV Network Architecture

The majority of BTV networks installed between 1996 and 2002 utilize Direct Video Broadcast (DVB) technology platforms. The DVB platform transmits video programming encoded using the MPEG-2 standard. MPEG-2 systems are capable of providing broadcast quality video; in fact,

MPEG-2 is the standard for encoding movies on DVD. (Its predecessor, MPEG-1, provided VHS quality video.)

The typical architecture of a conventional BTV network is illustrated in the following figure:

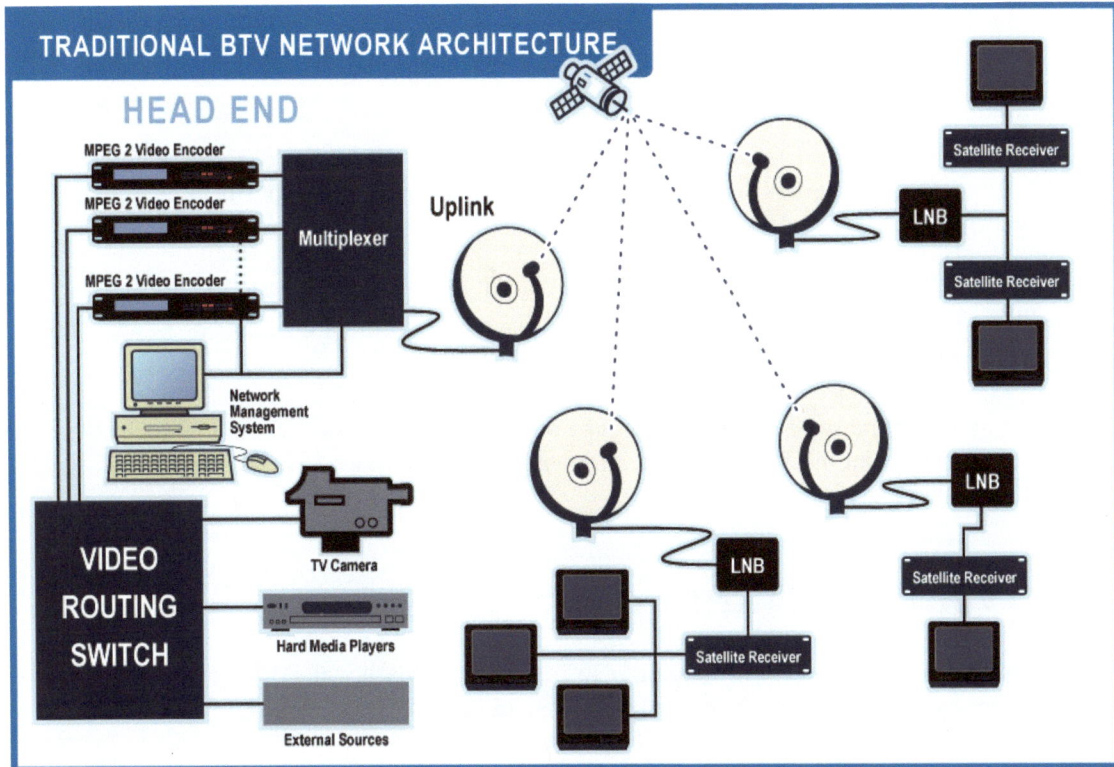

Figure #2: Traditional BTV Network Architecture

In a BTV network, analog source video from a live camera, tape or DVD playback source, or a remote feed, is routed or switched to a series of MPEG-2 video/audio encoders for digitization and compression. The number of MPEG-2 encoders is scalable, depending on the number of separate video channels that comprise the enterprise network. (Often, an additional MPEG-2 encoder provides redundancy.)

The digitized MPEG-2 video is coded and sent to a multiplexer, that merges video from multiple encoders into a DVB transport stream and sends it to the uplink for broadcast over the satellite.

The signal transported over the satellite is controlled by a network management system (NMS). Most BTV networks use the management system to control the encryption and scrambling of the DVB signal to ensure network security. The network management system also allows a network operator to authorize specific remote satellite receivers to decode the programming. It also tunes the satellite receiver to a specific program identifier (PID) for viewing of an individual program from the DVB signal.

At the receive location, the program signal is downlinked from the satellite by a parabolic antenna (typically 1.2 or 1.8 meters in diameter). The signal is then amplified and lowered in frequency by a Low Noise Blockconverter (LNB) located at the satellite antenna. The signal is then transported on a coaxial cable to the satellite receiver. This satellite receiver functions as an integrated receiver and decoder (IRD) of the satellite signal.

When authorized by the network management system, the IRD decodes a specifically assigned MPEG-2 video stream, and puts out an analog video signal for display on a television. Some BTV networks distribute programming to multiple viewing locations at one downlink site by sending the output of the IRD to an RF distribution network. The signal is then routed throughout an office building or a campus for viewing on multiple TVs in different areas such as an office, lobby, cafeteria, or conference rooms.

If viewers at a receive location need to watch multiple programs off the DVB carrier simultaneously, additional satellite IRDs are required. These IRDs receive their distinct input from a splitter placed on the output of the LNB.

Programs are recorded for later viewing on a VCR installed between the receiver and the TV. The VCR can be controlled either by local personnel or by the network management system, using a serial data connection.

Some enterprises also transport data over this type of network. This data often consists of large data files, which are sent from the IRD through an opportunistic data port, to a device such as a router, and then transmitted to a networked device.

3.2   BTV/IP Head End Architecture

The head end of a BTV/IP network is similar to that of a conventional BTV network; however, it contains additional components to leverage the full capabilities of the technology.

BTV/IP services are based on DVB transport over satellite. DVB transport specifies that all data packets in the transport stream comply with the MPEG-2 standard. However, BTV/IP services often involve the transport of non-MPEG-2 video and other IP data.

MPEG-2 video is broadcast quality and is intended for viewing on a TV. The lowest data rate feasible for MPEG-2 technology is approximately 1.5 Mbps. However, most BTV networks transmit MPEG-2 video at 2.5 Mbps or higher. To transmit data rates at or below 1.5 Mbps, other video encoding standards are used, such as MPEG-1, MPEG-4, and Windows Media 9.

Transport rates for streaming video on typical LANs must be below 1.0 Mbps to avoid network congestion and to be viewed on standard office PCs. These non-MPEG-2 video streams typically consist of IP packets that are converted to MPEG-2 for transmission over the satellite. This is achieved using a device called an IP Encapsulator (IPE). The IPE allows IP data packets, which are small, to be packaged inside larger MPEG-2/DVB packets. Packets are then multiplexed with other MPEG-2/DVB packets and uplinked to the satellite. They are then unpackaged at the designated receive sites as a stream of IP video packets.

Furthermore, IP data from any non-MPEG-2 encoder or from the head end LAN can be input to the IPE. This data can include archived IP video and multimedia, large data files, training materials and learning management system interconnectivity, and cached internet data.

The network management system for a BTV/IP head end must be more robust than that of conventional BTV network to handle the multiple tasks that it is designed to accomplish. To this end, enterprises should consider the implementation of multiple management systems at the head end to meet their specific requirements, such as:

- Broadcast network management
- Conditional access and network security
- Network monitoring
- Content distribution and management
- Program scheduling
- Playlist creation and management
- Stream creation and management

The functionality of these systems is included in the discussions of the various BTV/IP services in Section 3.14 *Network Management System Functionality*.

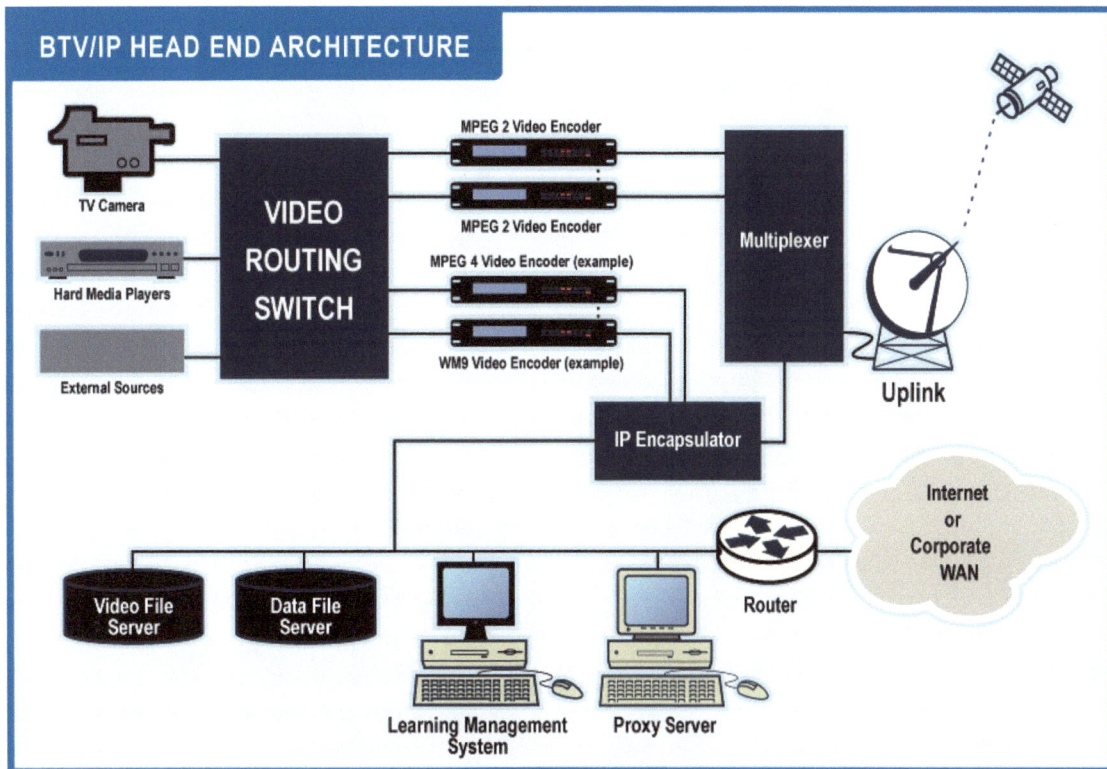

Figure #3: BTV/IP Head End Architecture

3.3 How a BTV/IP Media Gateway Works

The BTV/IP media gateway is the classification of IP satellite receiver that offers the widest range of BTV/IP services.

Figure #4 details the routing paths available in one such product, the Wegener iPump. Other BTV/IP media gateways contain similar routing paths.

The live MPEG-2/DVB carrier enters the BTV/IP media gateway, where the *DVB receiver* first processes it, and then descrambles and decrypts authorized programming. Next, the MPEG-2 streams are processed by *packet identifier filters* (PID filters), which separate the individual MPEG-2 programs from the multiplexed streams. The live video is then routed to a video decoder that converts the video into NTSC and outputs the receiver for display on a television.

These live programs may also be recorded on the hard drive for future VOD playback. Either the head end or the remote site may initiate this live recording.

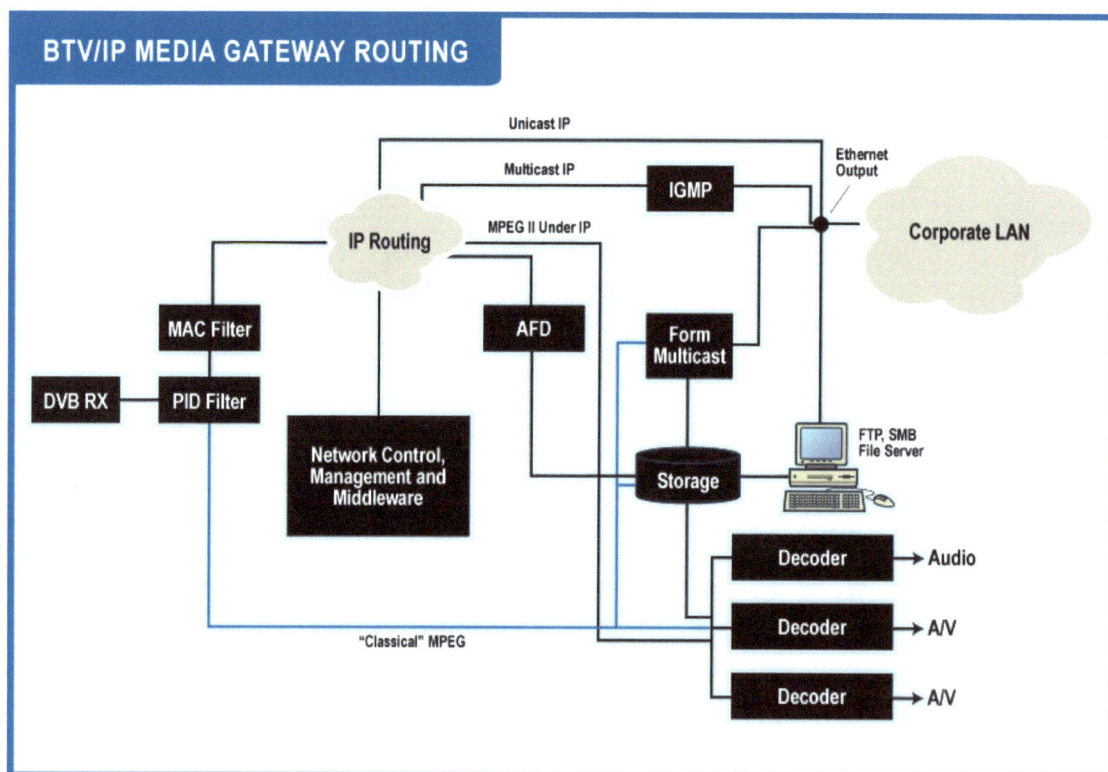

Figure #4: BTV/IP Media Gateway Block Diagram

Non-MPEG-2 video and data are processed by the *DVB receiver* and *PID filters* and sent to *MAC filters*. *MAC filters* look at the addresses of the incoming data and determine the specific device or port on the LAN to deliver the content. The video or data is then routed to its desired location.

Unicast traffic is sent directly onto the LAN to the targeted device. Multicast traffic is wrapped with the Internet Group Management Protocol (IGMP) and sent onto the LAN to any devices included in the multicast group. IP data may also be routed to the BTV/IP media gateway's *internal storage* using reliable file transfer (RFT).

IP routing is controlled by the network management software, which resides inside the BTV/IP media gateway. This software contains all of the middleware required to support BTV/IP services.

Stored content can then be served onto the LAN using protocols such as File Transfer Protocol (FTP) or multicast onto the LAN in a similar fashion as the live streams through the *Form Multicast* processing of the receiver.

3.4   Live Video over Satellite to a Television

Video content broadcast live over satellite to a television is achieved in the same manner for both BTV and BTV/IP networks, using the MPEG-2 standard for video content on the DVB platform. The head end and the required NMS functionality are the same in both types of networks, as are the antennas and remote site LNBs.

Of the three categories of BTV/IP receivers addressed in this Guide, only the BTV/IP media gateway (and traditional IRDs) possesses this capability. IP satellite routers and receivers lack the required video decoder.

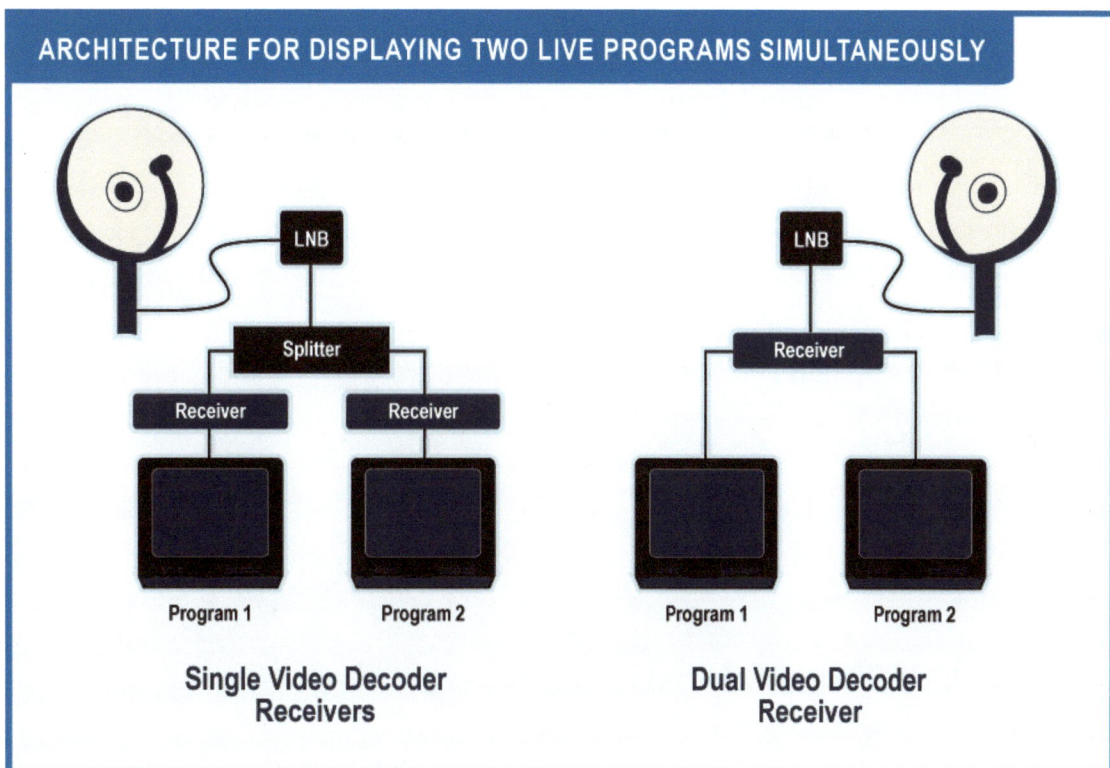

Figure #5: Architecture for Displaying Two Live Programs Simultaneously

One advantage of certain BTV/IP media gateways is the inclusion of a second video decoder. This allows a single satellite receiver to downlink and display two distinct programs simultaneously. This is not possible with IRDs or BTV/IP media gateways with only one video decoder. Refer to Figure #5.

For networks that must simultaneously receive multiple programs, this second channel feature offers a single box solution, without a significant increase in cost for each respective receive location.

Also note that the BTV/IP media gateways can be used for displaying both live and stored video content. An audience in one conference room can view a live broadcast, while a group in another room views an archived program off the same receiver.

3.5   Recording Live Video at the BTV/IP Media Gateway

The BTV/IP media gateway is the only receiver group featured in this Guide that contains a hard drive for storage of archived material. The advantage is that the video need only be broadcast once over the satellite. Once the content is stored, the assets can be remotely managed and administered by the NMS. An additional benefit offers the viewer at each location easy access to archived programming through the receiver middleware.

The network architecture for recording a live broadcast on the BTV/IP media gateway's hard drive requires no additional hardware. Software in the NMS gives the network administrator the ability to direct any or all remote media gateways to record a program.

Remote users also have the ability to direct the BTV/IP media gateway to record a program, either from the unit's front panel or through the network connection.

The BTV/IP media gateway has the ability to record a program while decoding it for viewing on a TV and/or streaming the program onto the LAN for viewing at employee desktops. Live programs can also be stored at remote locations. Media gateways can record multiple programs simultaneously. For example, it is possible for a remote site to play a live program to a TV screen while recording two other programs to the hard drive for later viewing.

3.6 Storage of Pre-Recorded Video and Other Files to Remote Sites

Pre-recorded video files can be sent to storage devices at remote sites using BTV/IP technology. These files are treated as data and can be transferred using the same satellite file transfer protocols as traditional data. This can be achieved using any of the three classes of BTV/IP receivers. The only difference between them would be the storage device at the remote site, which would be a media gateway or any other storage device networked on the remote LAN. These options are shown in Figure # 6.

These files can be unicast, multicast, or broadcast over satellite reaching one, some, or all of the sites in the network.

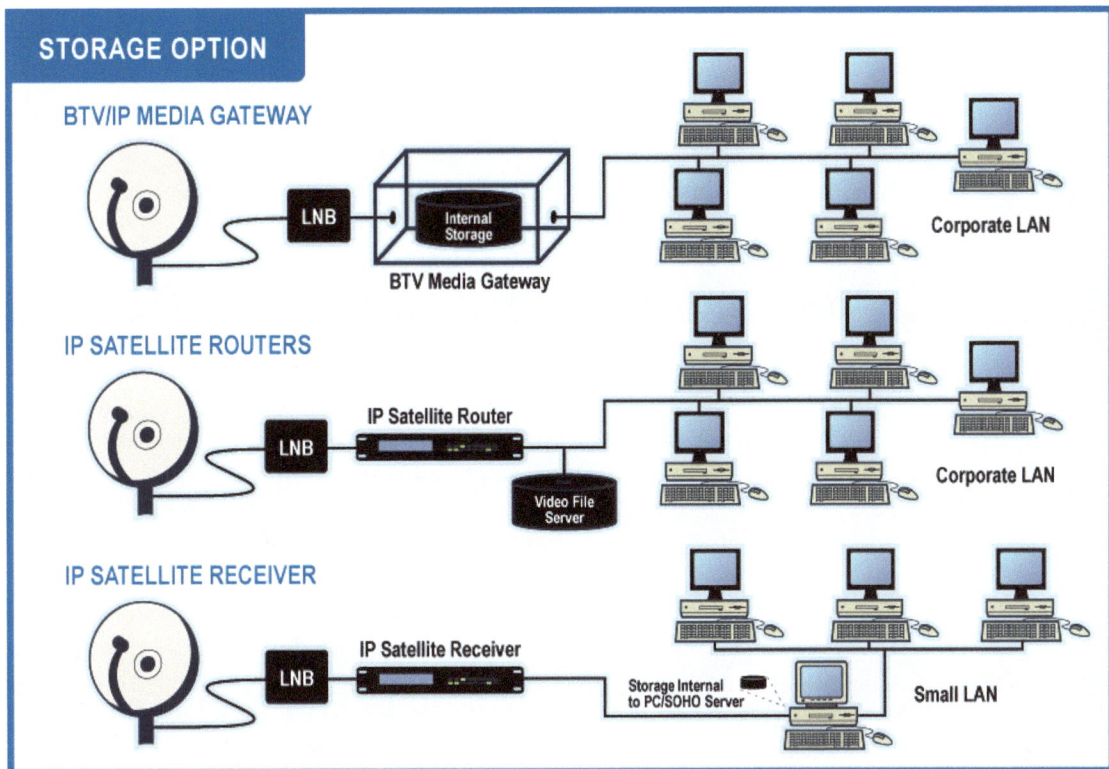

Figure #6: Storage Option

Large file transfer over satellite was one of the first IP multicast services to gain acceptance during the late 1990s. The functionality and the graphical user interfaces (GUI) are robust and reliable. The following are samples images of NMS interfaces for large file transfer. Companies such as Kencast provide software platforms for reliable file transfer (RFT) that can be used with receivers that do not provide NMS solutions.

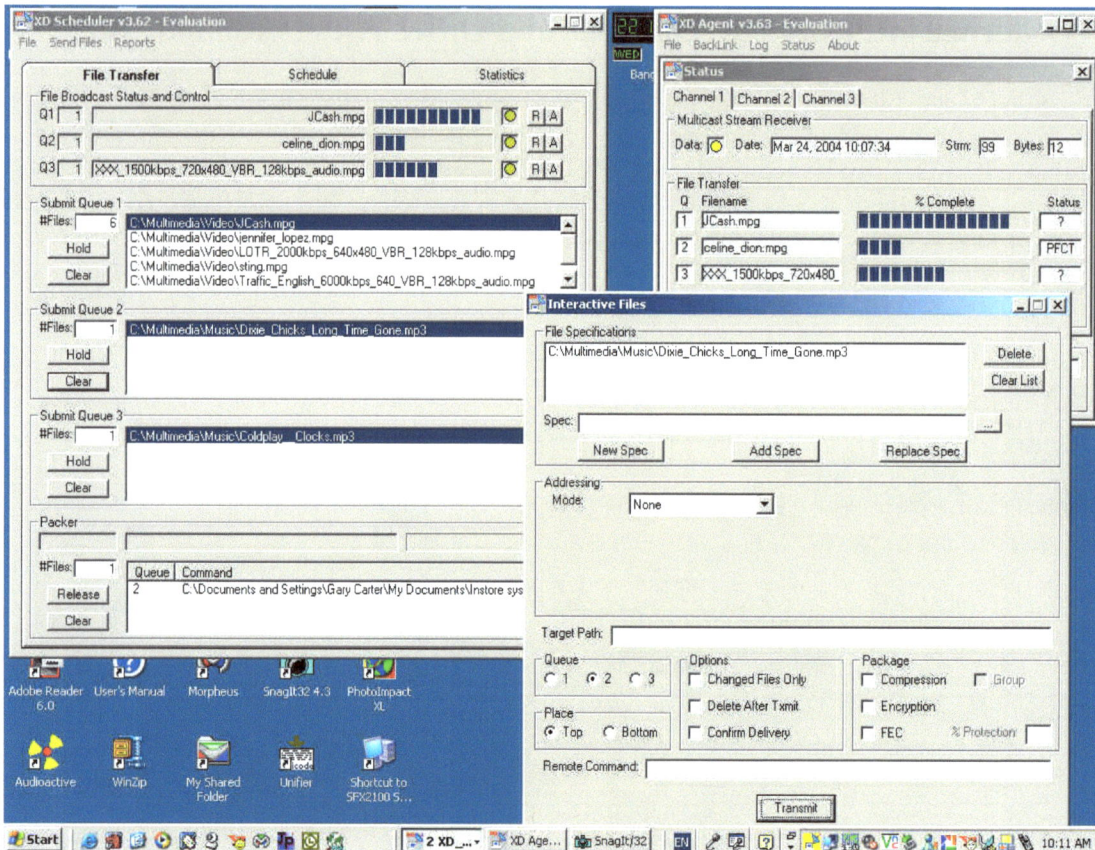

Figure #7: Screen Capture of File Transfer GUI

A common issue facing the enterprise considering BTV/IP network implementation is how much storage the network requires for video content. To answer this question, the enterprise must determine the number of hours of programming it expects to store at the remote sites and the data rates at which the video must be encoded for remote viewing. The following table estimates the number of hours of storage for video encoded at various data rates on hard drives of different capacities.

CONTENT STORAGE HOURS

MBPS RATE	DRIVE SIZE					
	40GB HOURS	80GB HOURS	120GB HOURS	160GB HOURS	200GB HOURS	240GB HOURS
1.5	56.3	115.6	174.8	234.1	293.3	352.6
2	42.2	86.7	131.1	175.6	220.0	264.4
2.5	33.8	69.3	104.9	140.4	176.0	211.6
3	28.1	57.8	87.4	117.0	146.7	176.3
3.5	24.1	49.5	74.9	100.3	125.7	151.1
4	21.1	43.3	65.6	87.8	110.0	132.2
4.5	18.8	38.5	58.3	78.0	97.8	117.5
5	16.9	34.7	52.4	70.2	88.0	105.8
5.5	15.4	31.5	47.7	63.8	80.0	96.2
6	14.1	28.9	43.7	58.5	73.3	88.1
6.5	13.0	26.7	40.3	54.0	67.7	81.4
7	12.1	24.8	37.5	50.2	62.9	75.6
7.5	11.3	23.1	35.0	46.8	58.7	70.5
8	10.6	21.7	32.8	43.9	55.0	66.1
8.5	9.9	20.4	30.8	41.3	51.8	62.2
9	9.4	19.3	29.1	39.0	48.9	58.8
9.5	8.9	18.2	27.6	37.0	46.3	55.7
10	8.4	17.3	26.2	35.1	44.0	52.9

Table #4 – Content Storage Hours

3.7   Live Video over Satellite to the Desktop

BTV/IP media gateways, IP satellite routers, and IP satellite receivers all have the capability to stream live video over satellite to the desktop or other networked display devices. This streaming media is often encoded at MPEG-1 quality or less, rather than MPEG-2 broadcast quality. This is simply due to the bandwidth constraints of most enterprise LANs.

Streaming media encoders, located at the head end, produce video in IP streams. These streams are processed by an IP encapsulator for transport over a DVB satellite carrier. Once processed by the remote satellite receiver, the video is sent as IP streams onto the network from the Ethernet port on the receiver. These streams can be unicast or multicast. The Internet Group Management Protocol (IGMP) is used for multicast traffic. This protocol is an extension to the Internet Protocol, used by IP hosts to report their host group memberships to immediately neighboring multicast routers.

All BTV/IP satellite receivers have the ability to transport multiple streams simultaneously onto the remote LAN. However, the number of streams varies by unit. Streams can be administered using some NMS systems.

27

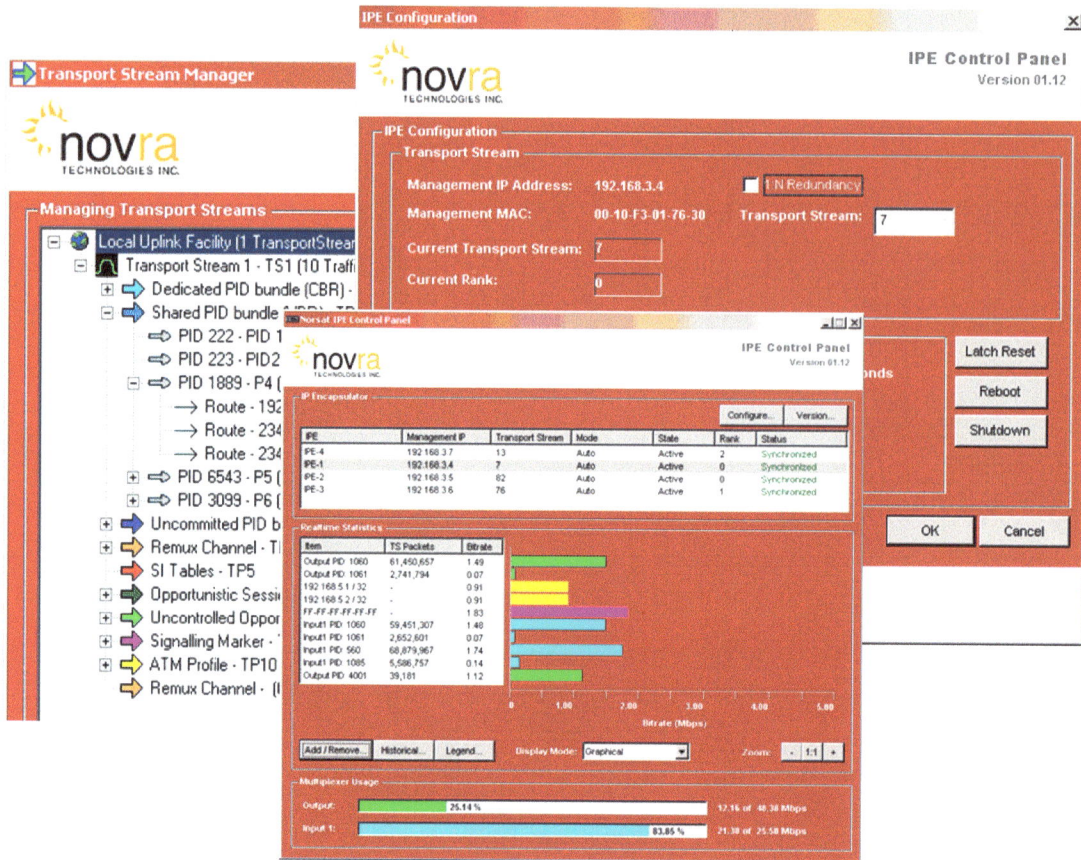

Figure #8: Screen Capture of Stream Management GUI

Users with networked connections can view a live stream by checking interfaces such as an Electronic Program Guide (EPG) and simply clicking on the desired stream. The typical EPG is similar to those provided on consumer cable and Direct to Home (DTH) network services, such as DirecTV and Dish Network.

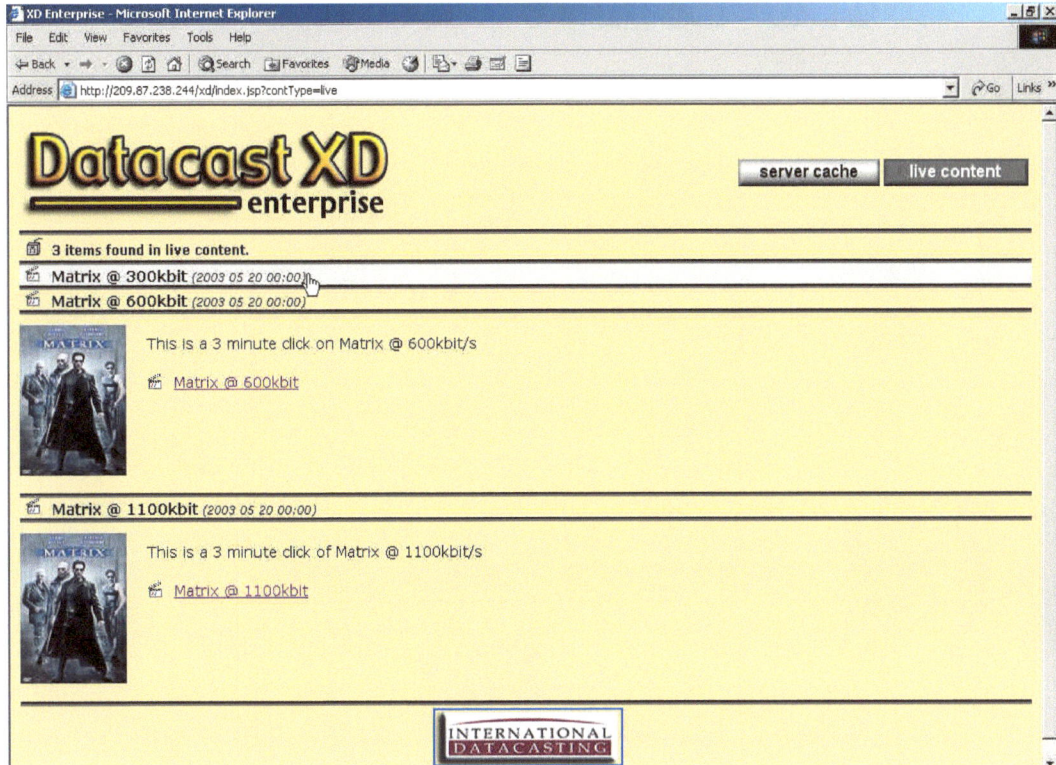

Figure #9: Screen Capture of Stream Select GUI

Media players are required on any PC that will view video streams. In most cases, the BTV/IP network is transparent to the encoder/decoder required (functions in the same manner as the Internet). The receiving desktop must have the necessary decoder to view the video. Some BTV/IP platforms have their own media players and encoders for these applications.

3.8 Video-on-Demand to a Television

Video-on-demand (VOD) programming can be played to a TV from a hard drive using a video decoder, which are built into devices such as media gateways and dedicated media players.

Dedicated media players are devices often found in retail broadcast networks, commonly referred to as personal video recorders (PVRs). These devices typically have Ethernet inputs, hard drives, and video decoders for playing video on a TV screen, selected from playlists downloaded to the device over the satellite network.

VOD programming stored on media gateways can be accessed by a networked connection. Some products use browser-type interfaces to access this information. Others use more robust EPGs that contain metadata about the programs to enhance the viewer's search process. The desired video file can then be selected for output to a TV screen. Refer to Figure #9 above.

Some viewers consider it inconvenient to use a LAN interface to access the content for display on a TV screen. Media gateway products are beginning to provide users with the ability to view the listings of stored video content from EPG displayed on a TV. The EPG is accessed using a hand-

29

held remote. Since the media gateway is not necessarily co-located with the TV, RF remotes are preferred over infrared remotes.

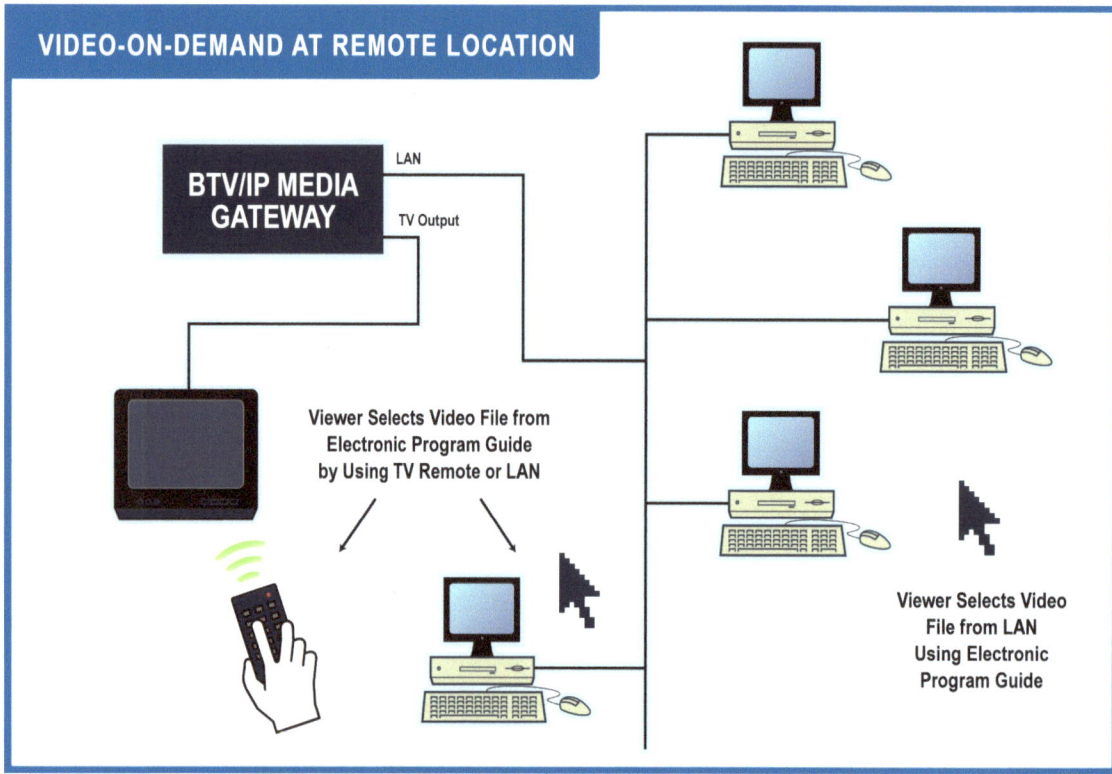

Figure #10: Video-on-Demand at Remote Location

Figure #11: Screen Capture of Program Selection

3.9 Video-on-Demand to the Desktop

Video-on-demand to the desktop is accomplished in much the same way as it is to the TV. Video files are typically accessed from either a media gateway or a networked storage device such as a video file server. Access is provided from either the browser or an electronic program guide. Kiosk applications work in much the same manner as VOD to the desktop.

3.10 Retail Network Applications

Retail network applications or digital signage, such as point-of-purchase (POP) displays, leverage the BTV/IP technology found in media gateways and network management systems to empower their networks.

The architecture of these networks varies depending on the services offered, and often includes additional devices such as dedicated media players. Displays for these applications also vary widely, ranging from small displays at gas pumps to large LCD displays in banks. The figure below shows a variety of network architectures for retail signage applications.

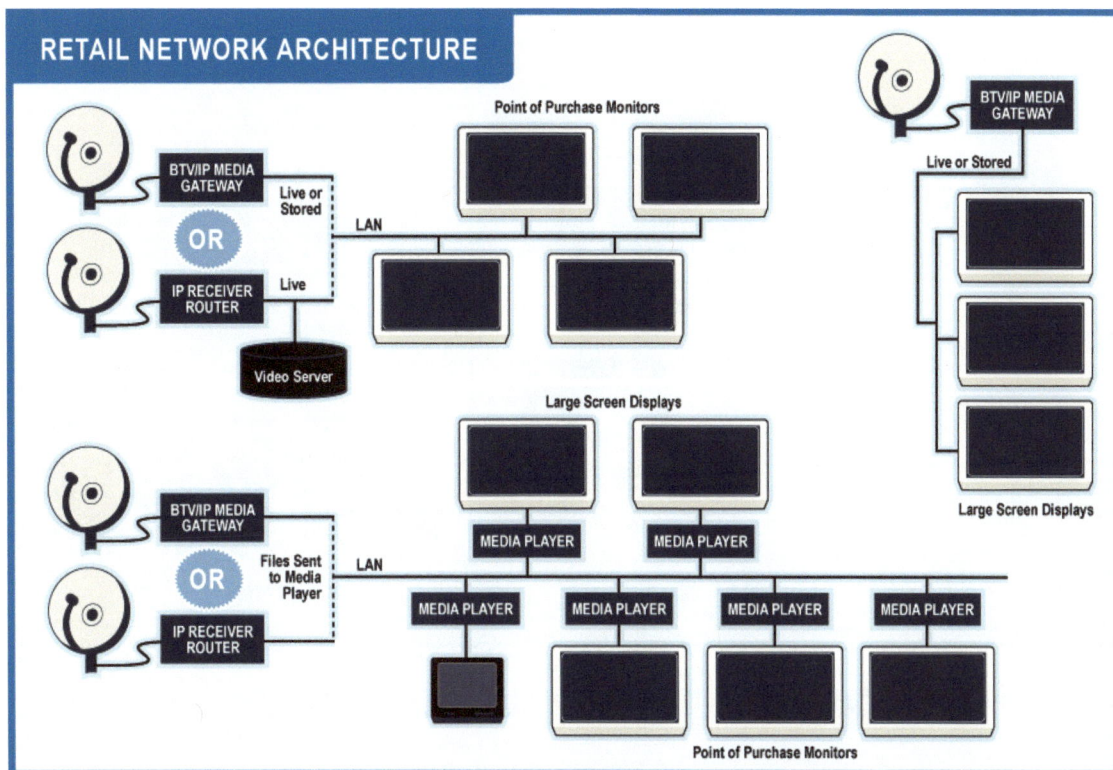

Figure #12: Retail Network Architecture

One of the key elements of these networks is the ability to multicast playlists of video content to the remote locations. A single media gateway may hold up to 100 separate playlists. These playlists can schedule stored programs for display based on complex rules intended to fine tune messaging. This functionality varies depending on the technology platform chosen. Also note that networks with dedicated media players may elect to use the playlist creation tools that accompany the player over those of the BTV/IP technology.

Because they are appliances connected to the network, media gateways and dedicated media players can report playback statistics to the head end NMS. Typically this data is sent as flat files on a terrestrial WAN and used to verify advertisement plays for billing purposes and to help measure the effectiveness of the content.

3.11   Interactive Distance Learning Applications

Numerous enterprises have leveraged BTV satellite networks to effectively provide training to associates in the field. Products such as the One Touch 5™, Arel VirtualClass™ or Savant Technology Group CADE™ enable student response and voice interaction. These interactive voice and data capabilities can be provided with a media gateway in the same manner as with a traditional IRD, as shown in Figure #13.

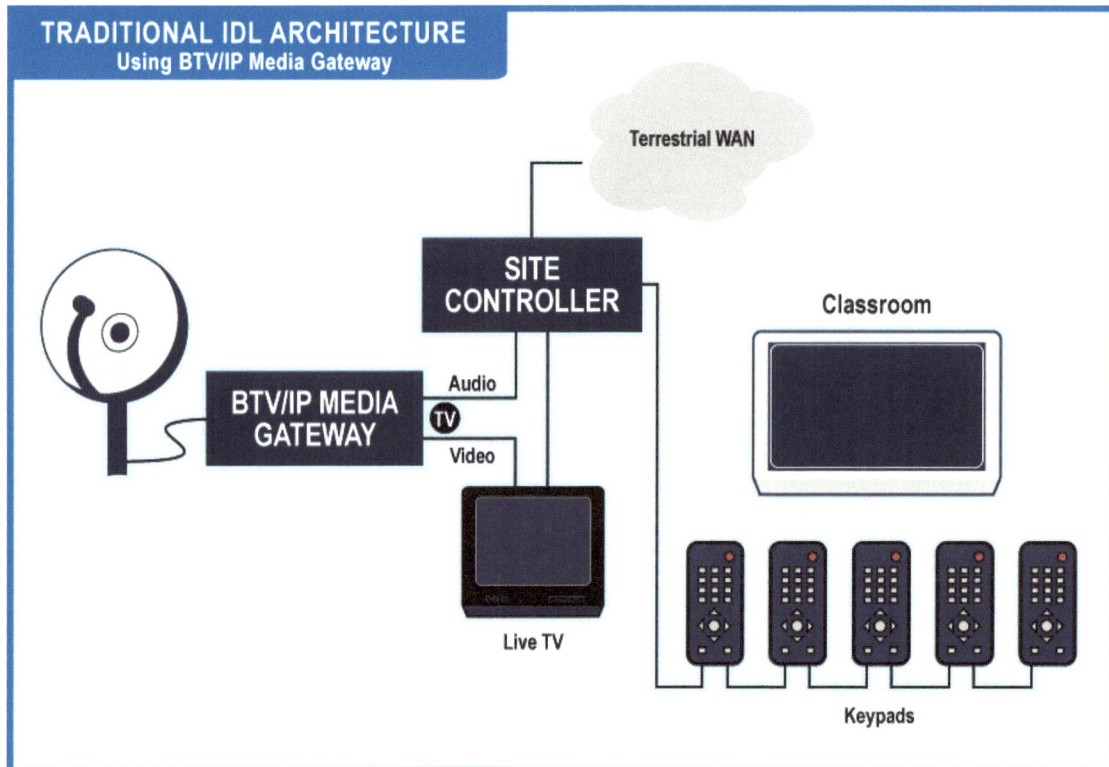

Figure #13:   Traditional IDL Architecture Using BTV/IP Media Gateway

BTV/IP provides the network architecture to support high bandwidth desktop training solutions in addition to the remote classrooms. In this architecture, the WAN bandwidth can be supplemented for multicast streams carrying training content to networked PCs as well as video monitors in the virtual classrooms. The student interaction from the PC or classroom would be handled through terrestrial connections such as the corporate WAN, the Internet, or even dial-up connections.

BTV/IP technology is also used for asynchronous self-paced learning. This application allows training materials to be delivered over satellite to remote storage. The student then accesses the training content from the server using a LAN connection and completes the coursework. Student progress and levels of achievement are tracked by the learning management system (LMS) and reported back to the main office.

Figure #14 shows how these computer-based solutions integrate into IDL networks.

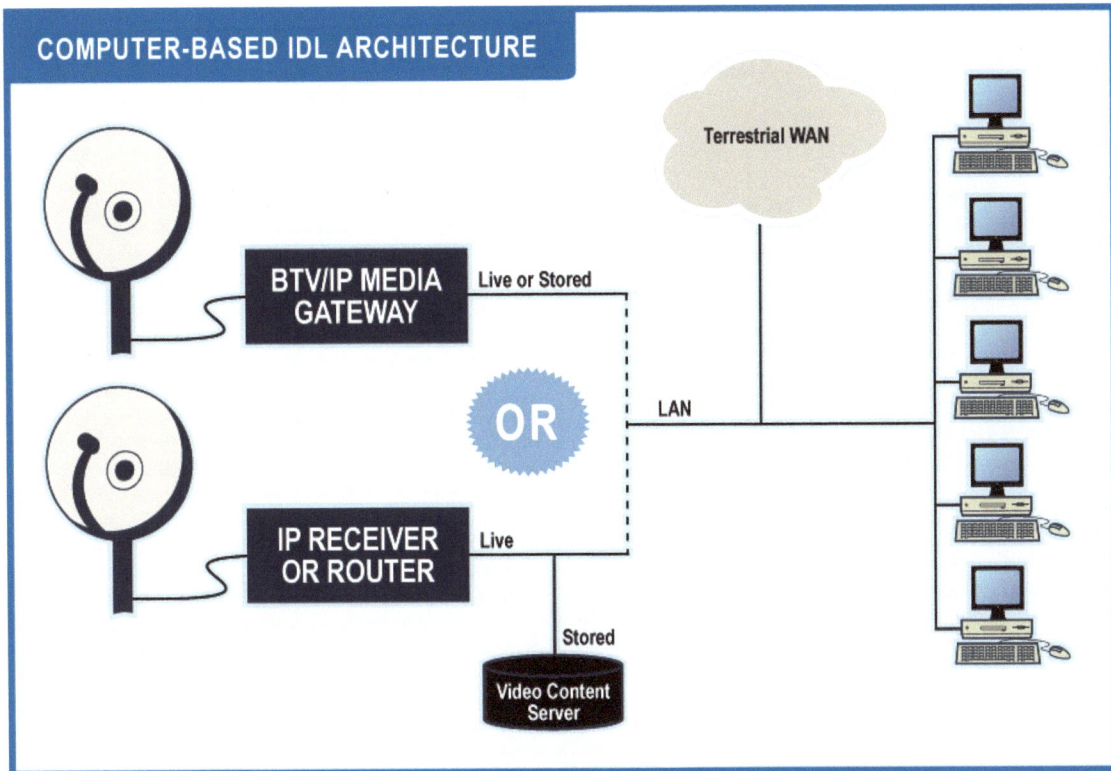

Figure #14: Computer-Based IDL Architecture

3.12 Satellite as an Overlay to the Terrestrial WAN

Terrestrial wide area networks (WANs) are often used to conduct mission critical services such as processing credit card transactions and transmitting financial data. The quality of service for these networks is often critical to the organization's success. Therefore, non-mission critical applications with high bandwidth demands may be unwelcome on the enterprise WAN. Video services often fall into this category.

Satellite offers an excellent supplementary, complementary solution to these traffic issues by serving as a high-bandwidth overlay to the terrestrial WAN. In this hybrid architecture, the terrestrial WAN continues to function as the primary network for information connectivity from the corporate headquarters to all remote sites (see Figure #15).

Non-mission critical information such as video – but also including a variety of data applications – are routed from the terrestrial WAN to the satellite uplink, where it is multicast to the appropriate locations over satellite. Once received at the remote site, the information is routed from the IP satellite gateway, router, or receiver onto the LAN and to its assigned location.

Return traffic, such as the acknowledgement of data receipt, is communicated over the terrestrial WAN, as are the initial data requests. The network management tools to achieve this are proven and reliable.

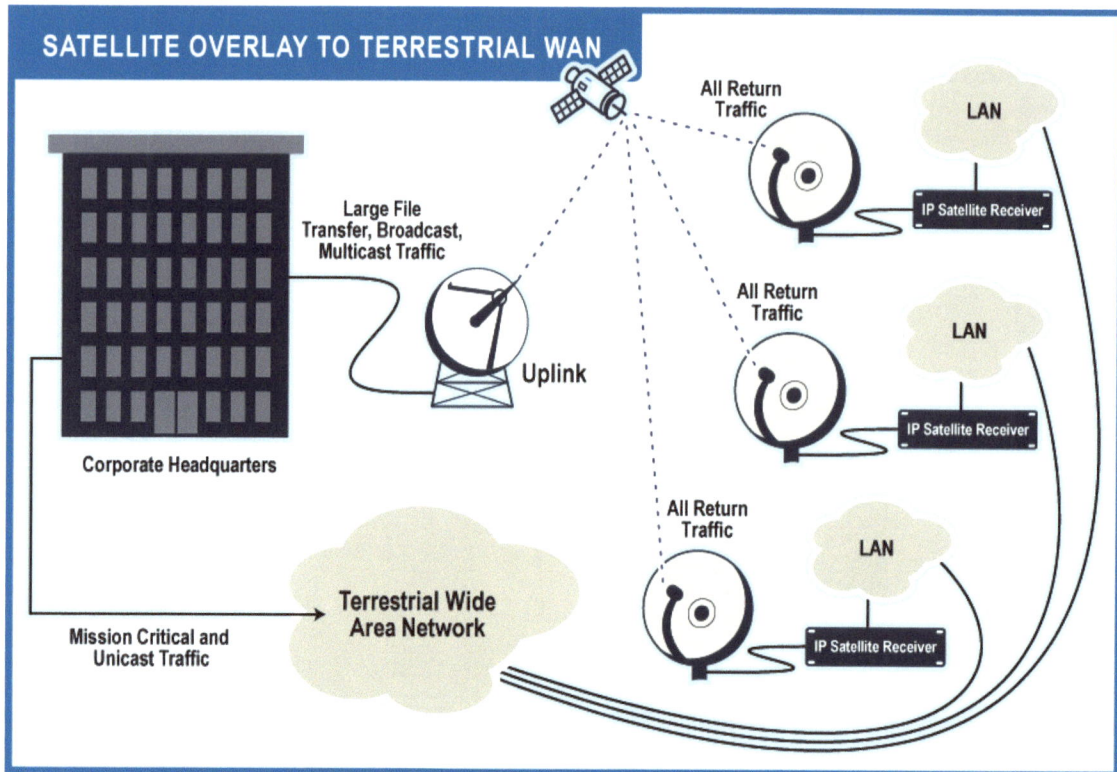

Figure #15: Satellite Overlay to Terrestrial WAN

Another advantage of this architecture is the ability to reach locations that are outside the reach of the terrestrial WAN. Simply put, high-bandwidth fiber optics is not ubiquitous, especially throughout rural areas. It may be years before these locations are serviced by terrestrial connectivity, if ever. In addition, enterprises may not find it to be cost effective to provide high-bandwidth terrestrial connectivity into small office home office (SOHO) locations. Again, a satellite overlay offers an excellent supplementary, complementary solution to this delivery challenge.

3.13   Hybrid Receiver Networks

A wide range of enterprise configurations, based on vertical industries such as pharmaceutical, grocery and retail stores, have different facilities, structures and communication requirements than do auto dealerships, hospitality organizations and financial institutions. As a result, viewing locations vary greatly in size, functionality, and mission. Examples include large district offices versus SOHO offices; retail stores with both public areas and behind-the-counter office and training space; manufacturing floors and offices; and financial institution lobby areas versus offices and conference rooms. Each of these environments may receive different BTV/IP services and require a different category of satellite receiver. As discussed in the *Executive Overview*, these hybrid satellite receiver networks can be configured using BTV/IP technology.

Figure #16 illustrates a network using the Skystream technology platform. This network is using all three classes of IP satellite receives/routers, which are controlled by the Skystream Z-Band network management system.

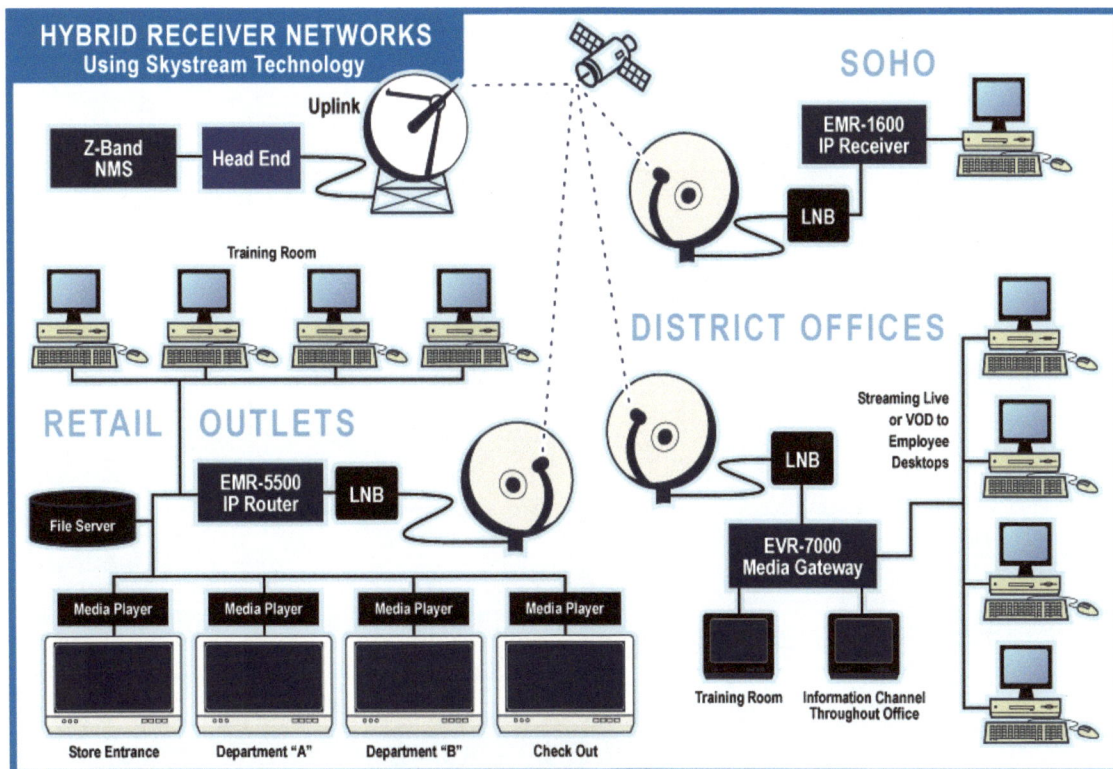

Figure #16: Hybrid Receiver Networks

In this example, home offices use the Skystream EMR-1600 IP Receiver to receive streams of corporate broadcasts as well as video files of relevant information for later viewing on the PC. The network also provides high-speed Internet access using satellite delivery to download information requested through a terrestrial connection.

District offices are using the EVR-7000 Media Gateway. These locations may have televisions in common areas for an employee information channel, while other TVs located in training rooms display both live and on-demand training programs. Corporate communications and training are also streamed to desktops, either live over satellite and/or from files stored on the media server.

Retail locations use the EMR-5500 IP Router. All files are stored on an external storage device. Associates receive training by accessing these files from PCs located in the training rooms. In addition, playlists are sent to media players located throughout the store, from which advertisements and promotions are played to large screen displays.

3.14 Network Management System Functionality

The functionality of the network management system (NMS) is one of the most critical elements of a BTV/IP network. NMS capability is as important a factor in choosing a technology platform as the satellite receiver itself.

Not all manufacturers offer an integrated NMS as part of a complete solution. These products rely on off-the-shelf software systems to provide functionality for applications such as file transfer.

Other elements of network management are provided by the DVB satellite technologies deployed at the head ends of such networks.

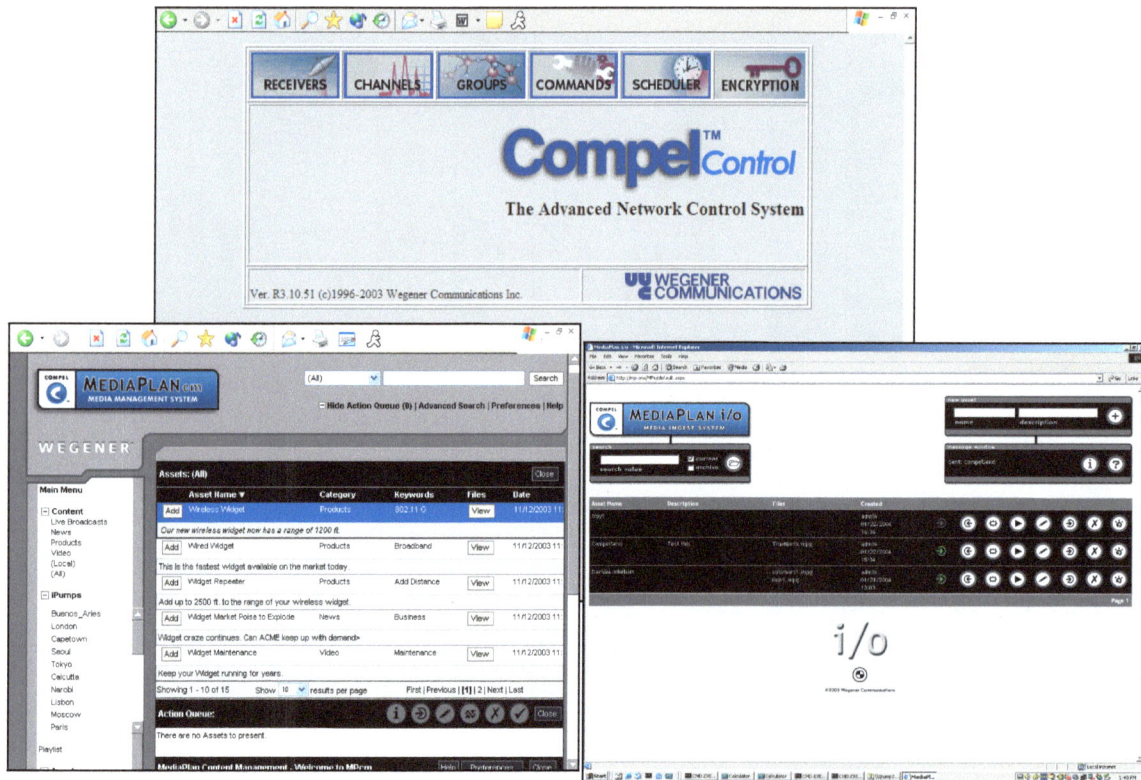

Figure #17: Screen Capture of Network Control System

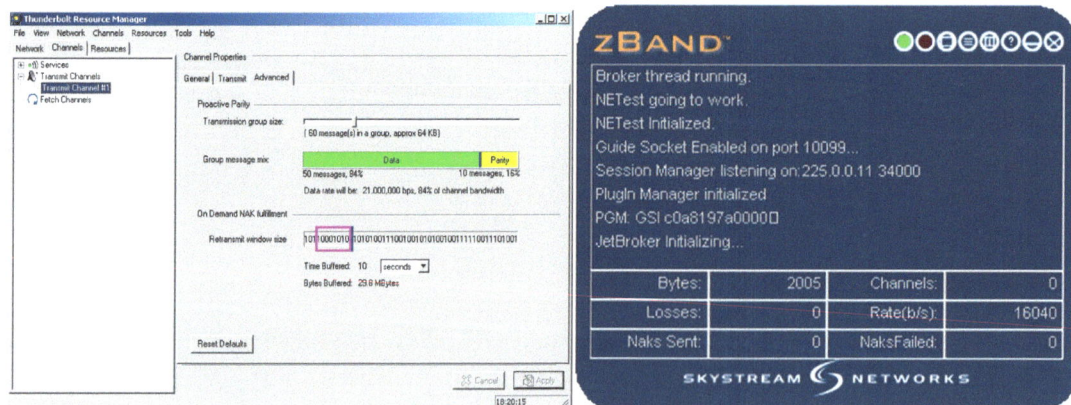

Figure #18: Screen Capture of Content Distribution

For many applications, these non-integrated NMS solutions are sufficient. Enterprises that have deployed networks using such architecture tend to be content and find no compelling reason to change their current NMS.

The levels of functionality amongst manufacturers included in this guide with network management systems vary widely. These differences are discussed in Section 5 *Product Comparisons of BTV/IP Satellite Receivers*.

3.15   Conclusion

The architecture of a BTV/IP network is defined by the services provided by each respective enterprise. The choice of a satellite-based BTV/IP receiver depends on the specific application for each network and the viewing objectives of each remote site. BTV/IP allows for different receivers using the same technology platform to be located throughout any given network. The NMS functionality ultimately defines how the services are implemented.

The table below details which BTV/IP services can be provided by BTV/IP media gateways, IP satellite routers, and IP satellite receivers.

BTV/IP SERVICES BY RECEIVER TYPE

	BTV/IP MEDIA GATEWAYS	IP SATELLITE ROUTERS	IP SATELLITE RECEIVERS
Live Video to a Television	YES	NO	NO
Recording of a Live Broadcast	YES	NO	NO
Video On Demand to a Television	YES	NO	NO
Live Video Streaming to the Desktop	YES	YES	SOHO
Video On Demand to the Desktop	YES	H/W	H/W
Digital Display	YES	H/W	H/W
Traditional Interactive Distance Learning	YES	NO	NO
PC-Based Interactive Distance Learning	YES	YES	SOHO
Large File Transfer	YES	YES	SOHO
High Speed Internet Downloading	NO	YES	SOHO

YES = Compliant   SOHO = Service compliant in Single PC or Small Office Only

NO = Non-compliant   H/W = Additional Hardware required to implement service

Table #5: BTV/IP Services by Receiver Type

4.0 IP Satellite Receiver Manufacturer Profiles

The manufacturers whose products are included in this report meet the criteria as providers of IP-based satellite receivers relevant to the BTV and distance learning industry. The criteria are:

- Their products meet the applications and needs of the enterprise user;
- They have units currently deployed in the field and/or established sales channels with industry service providers; and
- Their receivers are generally available through resellers and service providers.

Adtec Digital Inc.

Company Profile

Adtec Digital Inc. develops and markets audio-visual solutions for the delivery and staging of MPEG over analog or digital broadband and emerging IP networks.

Adtec's mission is "To develop and market high availability solutions that enable the global delivery and staging of MPEG over analog or digital broadband and emerging IP networks.

Adtec has delivered more than 27,000 MPEG digital video appliances working worldwide. In addition, Adtec provides audio/visual display and point-of-purchase (POP) solutions for the entertainment, lodging, and transportation industries, as well as theme parks and museums.

A detailed list of customers is on Adtec's web site, with a strong presence in a number of industries, including healthcare, broadcast and cable television.

Corporate Information	Founded in 1985 as Adtec Productions Incorporated, it is now Adtec Digital, with its headquarters in Nashville, TN. 42 employees **Privately Held** - Adtec is a privately held corporation with nearly all equity shareholders involved in the daily operations. Additionally, Adtec is debt free and maintains a positive cash flow.
Industry Experience	Adtec has design expertise in all aspects of embedded systems, including hardware design, real time operating systems, networking, and user interfaces. Its expertise is built on the fusion of hardware, firmware and software. Adtec's team implements integrated circuits, including high performance CPU's, specialized MPEG ASICs, and FPGAs. In the mid-1990's, Adtec designed an embedded solution that reliably played MPEG-1 without a PC: Soloist ™ Professional MPEG Player. Currently, over 2,000 of these units are still functioning on a daily basis. Since then their second generation MPEG 2 based servers have been shipped globally to thousands of customers in varying industries; to date more than 27,000 MPEG based video servers have been shipped.

Technology Partnerships	Acterna Callisto Media Systems Calypso Design Engage Communication	Hightower Systems Irdeto Access Coship Electronics MagicBox Medialon
Strategic Relationships (Sales Channels)	Globecast Business Television Video Jack Morton Worldwide Pittsburgh International Telecommunications	

Enterprise Customers (Partial Listing)	Alaska Channel, Inc. C.A.R.E. Channel CNN Airport Network First American National Bank Healing HealthCare Systems: The IBM Credit Corporation Kennedy Space Center St. Mary's	Medical Center Royal Caribbean International Safeway The University of Michigan Walt Disney Imagineering
Other Information	Adtec provides solutions for: ➤ Broadcast networks ➤ Telco Service Providers ➤ Cable Networks ➤ Enterprise Business Television Networks ➤ Educational Facilities, Medical Facilities, Entertainment and Security ➤ Professional A/V solutions for theme parks, museums, point-of-purchase advertising, hotel/motel, restaurants and welcome channels, medical patient and staff education, entertainment, transportation	
Contact Information	ADTEC Digital Inc. Sales/ Support/Administration 408 Russell Street Nashville, TV 37206 615/256-6619 Engineering/Manufacturing 2231 Corporate Square Blvd. Jacksonville, FL 32216-1921 USA 904-720-2003	sales@adtecinc.com www.adtecinc.com

Helius Incorporated

Company Profile

Helius develops business-class data broadcasting solutions with products that enable the simple, secure and reliable integration of satellite and terrestrial networks and managed data delivery. These solutions provide customers a comprehensive enterprise-wide data broadcast infrastructure for corporate broadcasts, distance learning and training programs, digital signage, software updates, database synchronizations and desktop Internet access.

Working with industry leaders in the global satellite space, Helius provides data broadcasting infrastructure for thousands of remote locations. Partners include providers such as GlobeCast, Broadcast International, Microspace, Globecomm, Convergent Media Systems, and BTV+. Partners use the Helius suite of products to complete their own solution offering for enterprise customers around the world.

The Helius portfolio of solutions includes IP-based transmission, reception, conversion and distribution technologies. These solutions deliver video, voice, application and file data across WANs, LANs and distributed desktops.

Corporate Information	Founded in 1995 with headquarters in Lindon, UT. 30 employees **Privately Held** - Helius is a holding under the Canopy Group investment family.
Industry Experience	Helius has provided satellite IP solutions software since its inception for companies such as Hughes Network Systems and Philips Electronics. Helius is using its current foundation of intellectual property as a springboard to further development and innovation. Helius has been awarded a patent entitled "Method and Systems for Asymmetric Satellite Communications for Local Area Networks." This patent applies to, but is not limited to, IP multicast routing technology to multiple clients over a Local Area Network, bi-directional electronic satellite communication supporting a variety of return channels, satellite-based Internet applications, satellite-based interactive solutions and other satellite-related applications. Helius' intellectual property portfolio includes other pending patent applications, copyrights to software products and programs, trade and industry secrets, and other rights in its proprietary technologies.
Technology Partnerships	Hughes Network Systems MediaMatics Technology Vultron Technologies OneTouch Systems

Strategic Relationships (Sales Channels)	**Domestic:** Broadcast International BitCentral CI Select Convergent Media Systems Diversified Media Group GlobeCast Globecomm Systems Hughes Network Systems Harris Corporation Keypoint Services Marshall Communications Microspace Communications OneTouch Systems Satellite Engineering Group StagePost York Telecom	**International:** BTV+ (Canadian Business Television) CEV HSS Inc. Inno Micro Corporation LTRT Quest Gencom Technology SpeedCast Telesat
Enterprise Customers	Flagstar Bank *(Digital signage, content distribution, streaming services)* Fujitsu *(Training, content distribution)* Pfizer *(Interactive training, content distribution, corporate communications, conditional access)* Merrill Lynch *(Streaming services & playback)* NTT Do Co Mo *(Streaming services)* NASA *(Steaming services, multicasting)* National School Fitness Foundation *(Digital signage, training, corporate communications)* Raytheon *(Streaming services)*	
Other Information	Helius provides solutions for: ➤ Satellite Service Providers ➤ Enterprise Business Television Networks ➤ Digital Signage Integrators ➤ Retail Service Providers ➤ Interactive Distance Learning Solutions	
Contact Information	Helius, Incorporated 333 South 520 West. Suite 330 Lindon, Utah 84042 U.S.A. 1-888-764-9020	info@helius.com sales@helius.com www.helius.com

International Datacasting Corporation (IDC)

Company Profile

International Datacasting Corporation (IDC) manufactures products, integrates systems, and provides service for broadband satellite and wireless networks in the multimedia content distribution market. IDC's products are designed and developed for enterprise networks, Internet infrastructure, and distance learning, as well as multimedia content distribution.

IDC's mission is "...to be the preferred supplier of digital wireless infrastructure solutions for broadband multimedia applications."

IDC's technology is used in U.S. domestic private networks, including Procter & Gamble, the American Association of Airport Executives, National Public Radio and PBS.

Corporate Information	Established in 1984. IDC's corporate headquarters (including R&D, manufacturing and teleport) are in Ottawa, Canada. Approximately 50 employees **Publicly Traded** on the Toronto Stock Exchange (TSX:IDC). Ownership of IDC is spread between Capital Alliance Ventures (CAVI, www.cavi.com), the Bank of Montreal Capital Corporation, employees, and shares traded on the open market. In 2002 IDC acquired Datacast Communications Inc., in 2002, adding a strong new software component to its asset base.
Industry Experience	IDC has partners and distributors in more than 40 countries worldwide, with more than 50,000 installations. IDC's strength in the distance learning market is its seniority, applications focus, and a broad international client base, including the Mexican Government's K-12 program, the Virtual University of ITESM with more than 1,500 receive sites across Latin America and major universities in China.

Technology Partnerships	Agilis Alcatel Bell Amino Bitcentral Callisto Data Planet Dyband Envivio Harmonic/Divicom Hauppauge Irdeto Access	Loral Skynet Marshall Communications Mayah Communications Microspace Communications Nagra Sigma Designs Spacebridge Tandberg Television Vipersat Networks Websat
Strategic Relationships (Sales Channels)	Bitcentral Grupo Etercom IGP Loral Skynet	Marshall Communications Microspace Communications Wancom Communications

Enterprise Customers (Partial Listing)	Domestic: American Association of Airport Executives (news and training) Loral Skynet (multicasting service) National Public Radio (NPR) (file-based content distribution) Boeing (digital cinema) GTE (weather news distribution) Dyncorp (weather news and defense applications) General Dynamics (distance learning and telemedicine) Public Broadcasting Service (PBS) (file-based multimedia content distribution) Clear Channel (digital audio) Procter & Gamble (corporate training) Raytheon (multicasting)	International: Reuters (news, financial data and photo distribution) Zhejiang University-China's second largest university, (distance learning) Telenor-financial information network Mexican Government's K-12 distance learning program Notimex news network Telefónica of Spain-financial data and weather info Virtual University of ITESM (across Latin America) – distance learning Canadian Forces Radio and Television service BUAP-telemedicine, Mexico and region
Other Information	IDC provides solutions for: ➤ Satellite service providers ➤ Enterprise Business Television Networks ➤ Digital cinema and telemedicine ➤ Internet Service Providers IDC is a member of the DVB organization, OCRI, and the National Association of Broadcasters.	
Contact Information	International Datacasting Corp. 2680 Queensview Drive Ottawa, Ontario, Canada K2B 8H6 613/596-4120	sales@intldata.ca www.intldata.ca

IPricot Broadband Solutions

Company Profile

IPricot is a manufacturer of broadband communication equipment for DVB-S IP (and video streaming) networks over satellite and wireless. IPricot provides smart networking devices for the delivery of data to end-users in a customized manner.

IPricot provides large corporations, governmental bodies, ISPs, network operators & integrators, VARs, Internet service & content providers and their clients with access to high-speed applications: Fast Internet, IP Multicasting, File Transfer, Streaming Audio / Video and Distance Learning.

IPricot's mission is to provide reliable, state of the art, user-friendly equipment to meet its customers' needs.

Corporate Information	Established in 1996 as DotCom SA, its name was changed to IPricot in 2001. IPricot North America Inc. was created in 1998. Corporate headquarters are located in Paris, France, with its North American headquarters in Montreal, Canada. 35 employees
Industry Experience	IPricot has been involved in the worldwide broadband market since 1995, developing interactive set top boxes in collaboration with Vivendi and Apple Corporation. IPricot 's core competency is its technical ability to design and integrate both its chipset and firmware into one advanced unit, which is then designed into next generation network & Internet appliances.

Technology Partnerships	Agilis Apple Corporation Irdeto Access Logic Innovations Newtec ND Satcom	Solectron Tandberg Thales Vivendi
Strategic Relationships (Sales Channels)	BitCentral Eutelsat GlobeCast IBM Global Services Marshall Communications Microspace Communications NewSkies Satellite	PanAmSat Polycom Satcom Resources Satellite Engineering Group Tandberg Television Valliance Networks
Enterprise Customers	Boeing Dyncorp General Dynamics Reuters	
Other Information	IPricot has international presence with headquarters in Europe and North America supported by business development and sales directors throughout the world.	

	IPricot provides solutions for: ➤ Satellite service providers ➤ Network operators & integrators ➤ Internet service & content providers ➤ Enterprise business television networks	
Contact Information	IPricot 3539 Saint Charles Blvd, suite 604 Kirkland, Quebec Canada H9H 3C4 Tel: +1-514-981-7308 IPricot SA 10-12 Avenue de Verdun La Garenne-Colombes F-92250, France +33-1-(0)-46-52-53-00	info@ipricot.com www.ipricot.com

Mainstream Data, Inc.

Company Profile

Mainstream Data, Inc. provides a broad range of services, equipment, and custom applications for the distribution, filtering, display, and collection of content: Real-time news, stock quotations, digital images, audio, multimedia files, training materials and applications.

Mainstream leverages satellite, Internet and wireless technologies—including many developed in-house—to deliver content to newspapers, banks, TV and radio stations, Web portals, retail establishments, and restaurants worldwide.

Mainstream has been involved in developing satellite technologies since 1987 and introduced its streaming Internet delivery service, MediasNet, in 1995.

Mainstream's mission is to simplify the process of creating, distributing, displaying, and managing multimedia content across the enterprise. The company provides tools and communications that enable companies to reliably deliver content to the boardroom or the desktop.

Corporate Information	Founded in 1985. < 100 employees Privately Held –The Carlyle Group, GE Equity, Investor AB, MCI WorldCom, New Enterprise Associates, The Wheatley Group, and others maintain ownership of Mainstream Data.
Industry Experience	With more than 18 years of experience as a turnkey network services provider, Mainstream provides equipment that is designed to be part of an integrated whole (although parts can also be purchased individually) for applications from broadcast networks to VSAT networks, DVB-S satellite receivers to hybrid multimedia IP distribution platforms; for companies looking to distribute corporate training or communications, memoranda, application updates or simply needing a reliable and cost-effective backup for their existing terrestrial networks. Mainstream offers a satellite footprint that covers North America, Europe, North Africa, and the Middle East.
Technology Partnerships	ViaSat, Inc.
Strategic Relationships (Sales Channels)	Mainstream uses its own direct sales force to reach potential customers.
Customers	Alcas Holding BV (Muzak Europe) Bloomberg DMX MUSIC IDC's ComStock NTN Communications The Associated Press Reuters Shenzhen Securities Communications Corporation (China) Terra Networks PR Newswire WeatherBank.

Other Information	Mainstream provides solutions for: ➤ Satellite service providers ➤ Enterprise business television networks ➤ Streaming Internet delivery ➤ Information Companies: Text news, streaming stock quotations & digital images ➤ Digital audio	
Contact Information	Mainstream Data, Inc. 375 Chipeta Way, Suite B Salt Lake City, UT USA 84108 801/584-2800	info@mainstreamdata.com www.mainstreamdata.com

Novra Technologies, Inc.

Company Profile

Novra Technologies is a broadband equipment company that offers products and services to transmit and receive IP over MPEG using satellite, cable and terrestrial communication links. This facilitates cost-effective distribution of data, video, audio, files and streams multimedia and Internet.

Corporate Information	Founded in November 2000. It currently has 25 employees. **Publicly Traded** - registered in Canada (NVI:TSX:Venture).
Industry Experience	Novra employs a seasoned management and engineering team with extensive experience in software application development, RF design, DVB, wireless systems development & deployment, and networking technologies. Novra continues to advance its NovraLink digital signage technology, based on continued experience from systems deployed since 2002.
Technology Partnerships	Many of Novra's customers become technology partners, with end customer requirements finding their way into Novra product.
Strategic Relationships (Sales Channels) Customers	Avanti Communications BTV+ Clear Channel Satellite Services EMS Technologies G2 Satellite Solutions Globecomm Systems Inc. Broadband Technologies Inc. HNS, Mediacorp IsatPlatform	Loral Marshall NGIT Novanet OrientSat Networks PanAmSat Planetary Data Weather Underground
Other Information	Novra provides solutions for: ➤ Satellite Operators ➤ Business Television Service Providers ➤ Systems Integrators	
Contact Information	1100-330 St. Mary Avenue Winnipeg, Manitoba Canada R3C 3Z5 204/989-4724	psheedy@novra.com www.novra.com

SkyStream Networks

Company Profile

SkyStream Networks is a provider of IP networking solutions for service providers, carriers and satellite companies to deliver digital media services to businesses.

Applications supported include satellite broadband delivery, media asset distribution, broadband entertainment portals, content delivery services for consumers and enterprises, video on demand Turbo Internet, Internet delivery over DTV, corporate communications, and distance learning.

SkyStream works with and through more than 150 service providers.

Corporate Information	Founded in 1996 with headquarters in Sunnyvale, CA. 120 employees Privately Held - Funding from 3i US, AOL Time Warner Investments, Comcast Interactive Capital, Amerindo Investment Advisors, Crosslink Capital, Granite Global Ventures, IVP, Integral Capital Partners, Mayfield, Norwest Venture Partners, WestBridge Ventures and Shaw Communications.
Industry Experience	Skystream has eight U.S. patents including several in the area of bandwidth optimization for delivery of next-generation video and data services.

Technology Partnerships	Corecess Global Inc. Hewlett-Packard Irdeto Access Lucent Technologies MicrosoftTV Motorola Nagra nCube NDS Group	OneTouch Systems Pace Pinnacle Systems RealNetworks SeaChange International Sigma Designs Stellar One Corporation Tandberg Television Vipersat Networks
Strategic Relationships (Sales Channels)	ND SatCom Hewlett-Packard Imagine Broadband British Telecom Broadcast Services Clear Channel Convergent Media Systems EchoStar Communications Eutelsat Gilat Satellite Networks GlobeCast Harris Corporation Intelsat	Loral Cyberstar Marshall Communications Microspace Communications Pathfire Pinnacle NewSkies Satellite Starband/Gilat Verestar ViaSat
Enterprise Customers	Exam Room Network Reuters Safeway Inc.	

Other Information	Sales and technical support offices are located worldwide, including: London, Paris, Hong Kong, Beijing, Switzerland and Seoul. SkyStream provides solutions for: ➤ Satellite service providers ➤ Enterprise business television networks ➤ Satellite broadband delivery ➤ Broadband networks ➤ Content delivery services for consumers	
Contact Information	SkyStream Networks 455 DeGuigne Drive Sunnyvale, CA 94085-3890 408/616-3300	info@skystream.com www.skystream.com

Wegener Communications

Company Profile

Wegener is a leading international provider of reliable broadcast quality products for data, video, and audio networks providing either Live or Store & Forward services.

Wegener provides products to major cable television systems and broadcast networks in the United States, Canada, Europe and Asia. In their 26-year history, Wegener has supported radio networks worldwide, satellite paging services, and is a leading provider for the US Horse Racing industry. In addition, Wegener is the leading provider of receivers and systems to the business music industry.

Wegener is ISO-9001:2000 certified and has deployed over 50,000 video receivers for broadcast television, business television and educational networks.

Wegener is "in the business of providing innovative communication products." Their mission is "total customer satisfaction."

Corporate Information	Founded in 1978, with headquarters in Atlanta, GA. Approximately 90 employees **Publicly Traded** (Nasdaq: WGNR)
Industry Experience	Wegener Communications began by producing audio and data products for the cable television industry, including MTV. This led to a relationship with Muzak in 1982 and Fox Television in 1991. Both of these relationships continue today. Over 75 networks use Wegener control systems to control over a 150,000 sites. Its Satellite Network Control, patented in 1989, has been a key part of Wegener's product line. Both Glenayre and Motorola integrated Wegener Satellite systems into their paging network systems. Wegener has provided addressable network control systems since 1989. COMPEL™, Wegener's network control system, provides networks with the ability to regionalize programming and commercials through total receiver control. COMPEL network control capability is integrated into Wegener digital satellite receivers.

Technology Partnerships	Scientific Atlanta Harmonic Lightwave, Inc. Irdeto, Inc.	TBC Integration Kencast, Inc.
Strategic Relationships (Sales Channels)	Globecomm Systems Ascent Media Network Services Thales Broadcast & Multimedia Satellite Engineering Group	Microspace Communications Mega Hertz TeraCom, Sweden Spacecom Systems
Enterprise Customers	Muzak Channel One Dow Jones Investor Network Eli Lilly and Company GE Medical Scientific Games Roberts Communications	

Other Information	Wegener provides solutions for:	
	➢ Store and Forward Media Delivery ➢ Enterprise Business Television networks ➢ Network Control Systems ➢ Broadcast television networks ➢ Cable television systems ➢ Radio networks ➢ Business Music Industry ➢ Satellite service providers ➢ Paging systems	
Contact Information	Wegener Communications 11350 Technology Circle Duluth, GA 30097 770-814-4000	mailto:jaybatista@wegener.com www.wegener.com

5.0   Product Comparisons of BTV/IP Satellite Receivers

5.1   Overview

This section of the Guide compares the products that have met the requirements for inclusion discussed in Section 1.7. These products have been broken into three categories: *BTV/IP Media Gateways*, *IP Satellite Routers*, and *IP Satellite Receivers* based upon the likely applications they would serve in an enterprise environment. The product comparisons are presented separately for each of the three categories. Table #5 shows the typical services that can be provided by each of these categories.

BTV/IP SERVICES BY RECEIVER TYPE

	BTV/IP MEDIA GATEWAYS	IP SATELLITE ROUTERS	IP SATELLITE RECEIVERS
Live Video to a Television	YES	NO	NO
Recording of a Live Broadcast	YES	NO	NO
Video On Demand to a Television	YES	NO	NO
Live Video Streaming to the Desktop	YES	YES	SOHO
Video On Demand to the Desktop	YES	H/W	H/W
Digital Display	YES	H/W	H/W
Traditional Interactive Distance Learning	YES	NO	NO
PC-Based Interactive Distance Learning	YES	YES	SOHO
Large File Transfer	YES	YES	SOHO
High Speed Internet Downloading	NO	YES	SOHO

YES = Compliant   SOHO = Service compliant in Single PC or Small Office Only

NO = Non-compliant   H/W = Additional Hardware required to implement service

Table #5: BTV/IP Services by Receiver Type

Several of the products included in this guide do not necessarily meet the exact definition of any of these categories. These products have been placed in the categories deemed appropriate.

5.2   BTV/IP Media Gateways

A BTV/IP media gateway is a one-way satellite receiver that has the ability to playback analog video formats, such as NTSC, to a television monitor and provides an IP network interface with router-type functionality capable of providing IP video streaming onto an enterprise LAN. In addition, a BTV/IP media gateway contains storage for video content along with the necessary middleware to enable specific BTV applications.

BTV/IP Media Gateways are different than BTV IRDs with an IP port. At the time of printing this Guide, there are several products available on the market that fit the latter description. These products are not included in this report. Some of these manufacturers indicate that their products will include BTV/IP functionality in future releases and may be included in future releases of this Guide.

The BTV/IP media gateway is likely to be the predominant technology for next generation business television networks. These products provide an "all-in-one" solution that could meet the requirements of the typical enterprise. As such, these products will be explored in greater depth than IP Satellite Routers & Receivers in this guide.

5.2.1 BTV/IP Media Gateway Product Overviews

The BTV/IP Media Gateways products included in this guide are the Adtec Edje-L, Helius 2500-S, Helius 1500-S, International Datacasting SFX2100, Novra SSP-100, Skystream EVR-7000, and Wegener iPump.

This section provides overviews of these products.

5.2.1.1 Adtec Edje-L

The Edje-L, manufactured by Adtec, is a MPEG digital video player and network appliance with a built in QPSK satellite receiver, and is designed for incorporation with satellite multimedia networks. Although this product differs in some respects from other BTV/IP media gateways covered in this guide, the Edje-L meets all of the requirements for inclusion in this field has an established customer base.

Adtec is a pioneer in embedded video player solutions. In the early days of IP-based video, the operating system associated with most digital video players and PC-based satellite receivers (typically Windows) regularly crashed requiring the systems to be manually rebooted. Adtec overcame this by developing its own proprietary platform, Video Transport Operating System (VTOS). This platform is highly stable and enjoys a reputation for rarely needing to be rebooted.

The drawback of the VTOS platform is that it does not support some of the user-friendly middleware applications common to other BTV/IP media gateways, such as a web-browser interface.

The Edje-L distinguishes itself with the quality of its support of items such as close captioning, its ability to seamlessly splice between video files from a playlist, and its reliability. For that reason, Adtec's Edje product line is found extensively in dynamic digital signage applications.

In 2003, the Edje-L was recognized with the National Association of Broadcaster's Award for Innovation in Media.

5.2.1.2 Helius 2500-S

The Helius MediaGate Router 2500-S is the only product to combine three critical functions into one simple, secure and reliable device. The 2500-S combines a DVB satellite receiver, IP compliant router, and open application server platform. The 2500-S delivers powerful IRD functionality including optional support for multiple satellite receiver cards in a single receiver. This product is capable of providing a full suite of BTV/IP services, such as DHCP, FTP, SMTP and more, and has great flexibility in how it can be customized to meet specific network requirements. The 2500-S can perform routing and application services simultaneously while also delivering full feature audio/video outputs to a TV or other display device.

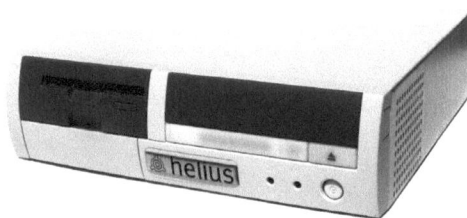

The strength of this product is in its ability to run multiple applications on the remote LAN. This is different from of the other BTV/IP media gateways included in this Guide that have their strengths in the satellite receiver and audio/visual interfaces. The 2500-S is a multi-purpose device rather than just a traditional IRD. This flexibility enables increased ROI at the remote site. By performing multiple functions the 2500-S allows customer to combine applications and minimize the hardware required at the remote site. The 2500-S supports add-on components such as Motion Detectors, Power Monitors and other components that are critical to Digital Signage implementations. This flexibility is further demonstrated in the ability to create custom reporting and feedback based on activity at the router. This information can be returned to the home office through FTP, SMTP or other transport options.

Another significant benefit of the 2500-S is the open platform, which not only enables application development and deployment, but also provides a simple and robust upgrade path. Using standard hardware components that can be field upgraded allows for increases in processor, hard drive, RAM and other key components. The Helius Media Core software which runs the 2500-S is also upgradeable and can be changed through online patches.

The 2500-S has been successfully deployed in both large enterprise and SOHO environments. Enterprises using the 2500-S like the product's ability to be customized to meet unique service requirements and acceptance by IT organizations as critical factors in the selection of this product.

5.2.1.3 Helius 1500-S

The Helius MediaGate Router 1500-S was developed in response to customer requests for an entry-level version of the 2500-S. The key difference between the two products is that analog audio/video output has been disabled on the 1500-S. The 1500-S allows customers to take advantage of the benefits of BTV/IP at an entry-level price point. Because the 1500-S has a hard drive and the product can be software upgraded to include audio/video output and to enable other features, it is included in the BTV/IP media gateway section, as opposed to the IP satellite router section.

The hard drive and advanced middleware options give this product robust capabilities to provide services such as VOD to the desktop that are not possible in a "single box" IP satellite router solution. The 1500-S represents a simple migration path for those customers who are interested in BTV/IP and need a path that allows minimal investment and a broad upgrade roadmap.

The MediaGate Router 1500-S has only recently been introduced to the marketplace and has garnered a good deal of attention. It provides an excellent addition to the Helius product line.

5.2.1.4 International Datacasting SFX2100

The SFX2100 is a fully functional BTV/IP Media Gateway manufactured by International Datacasting. This product differentiates itself from its competition in many key areas, most notably the quality of its components and design.

Key components of the SFX2100 are dual Ethernet ports allowing IT managers greater flexibility in how they implement IP-based video services. Four USB ports provide for greater options in implementing digital signage systems. It contains a superior satellite receiver capable of receiving 8PSK & 16QAM modulated signals, and an Apache Web Server for robust viewer interaction.

Another key component is the satellite receiver, which is able to receive carriers at 256 kbps – lower than any other product in this guide. This allows enterprises with small datacasting requirements to operate at lower satellite bandwidths, thereby reducing network costs.

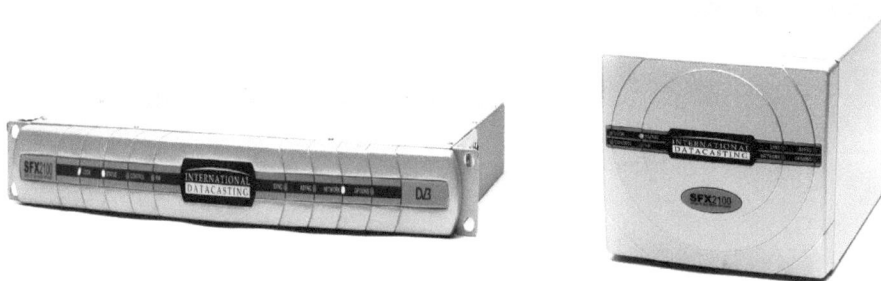

International Datacasting also provides its own content management system, Datacast XD. This system allows flexibility in conditional access and forward error correction schemes. It also allows control over return channels, event scheduling, and stream management.

5.2.1.5 Novra SSP-100

The Novra SSP-100 is a broadband media player that meets the criteria to be included in the BTV/IP media gateway category. The SSP-100 is developed and marketed toward retail communication networks and digital signage applications, although its capabilities reach into other BTV/IP services.

The SSP-100 differentiates itself by offering a large number of media formats for services such as in-store display. These media formats include MPEG-1, 2, & 4 video, MP3 audio, Quicktime, Flash, PowerPoint, Windows Media, and AVI files. Other features include default clip designation assuring that a proper video clip is following a system error and automatic restart on power cycle.

The SSP-100 operates on the Windows XP operating system. All of the other products in this category operate on either Linux or proprietary operating systems. One benefit of this configuration is that the SSP-100 can easily play video clips encoded in Windows Media 9.

The SSP-100 represents the playout component of Novra's *Novralink* media distribution system. *Novralink* software is designed for retail communication network services.

5.2.1.6 SkyStream EVR-7000

The SkyStream EVR-7000 is perhaps the most recognized BTV/IP media gateway product among BTV enterprise network operators. This product was one of the first to be integrated into the service offerings of BTV service providers and, as a result, has been deployed in several high profile BTV/IP networks.

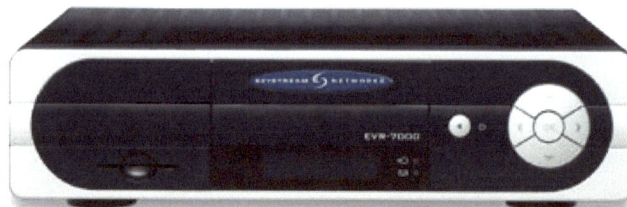

The EVR-7000 offers the full complement of BTV/IP capabilities. It is a video & multimedia complement to the EMR-5500, Skystream's widely accepted IP Satellite Router.

SkyStream's Network Management System, Z-Band, is reliable and robust, removing integration barriers that may otherwise impede network performance and perception.

5.2.1.7 Wegener iPump 6400

Wegener's iPump is a BTV/IP media gateway capable of supporting any BTV/IP service included in this Guide. This product is designed for use by broadcasters and cable television operators, in addition to enterprises. As a result, the quality of the audio and video interfaces and the performance in playback of video files is a major differentiator for this product.

Another key differentiator of the iPump is the inclusion of a second decoder. The iPump is the only product included in this guide that can play to live satellite broadcasts to separate televisions simultaneously.

Wegener has been the supplier of satellite receivers to the in-store audio company, MUZAK, for over 20 years. This has provided Wegener with significant experience in content management and conditional access systems. The result is the COMPEL™ network management system. The capabilities of COMPEL™ give network administrators maximum flexibility and control of their content as it is distributed to, and utilized at, remote sites across all BTV/IP applications. Many BTV service providers consider the COMPEL™ system to be the best in class.

5.2.2 BTV/IP Media Gateway Product Comparisons

This section of the Guide provides comparisons of the BTV/IP media gateways. In order to best accomplish this, the products are compared in nine categories, including:

- Audio/Video Interfaces
- Television Standards
- Storage
- Computer Interfaces
- Control & Functionality
- Operating Platform
- TCP/IP Support
- Security
- Receiver Specifications

A table is provided for each of these categories comparing the products against key parameters. Each parameter is defined, noting its significance the enterprise. Lastly, notes are provided on the product, highlighting key points relevant to specific category.

5.2.2.1 Audio/Video Interfaces for BTV/IP Media Gateways

BTV/IP MEDIA GATEWAYS > AUDIO VISUAL INTERFACES

	ADTEC EDJE-L	HELIUS 2500-S	HELIUS 1500-S	IDC SFX2100	NOVRA SSP 100	SKYSTREAM EVR-7000	WEGENER iPUMP
Composite Video	BNC	Yes	Optional	Yes	Yes	Yes	Yes
Component Video	Yes	Optional	Optional	Yes (2 optional)	No	No	No
S Video	Yes	Yes	Optional	Optional	Yes	Yes	Yes
Modulated Video	No	Optional	Optional	No	No	Optional Chan 3 & 4	Yes
Digital Video	Optional	Optional	Optional	Optional	No	No	Optional
Analog Audio	RCA phono & XLR	Stereo Miniplug	Optional	L & R RCA phono	Yes	L & R RCA phono	RCA (XLR optional)
Digital Audio	No	Optional	Optional	Optional	No	No	Yes

Table #6: BTV/IP Media Gateways – Audio Visual Interfaces

The seven manufacturers of BTV/IP satellite receivers detailed in this Guide each takes a slightly different approach to the Audio/Video Interfaces provided with their products. These differences are due to variations in the market segments targeted by each manufacturer.

The video Interfaces for these products consist of the following five types: composite video, component video, S-video, RF modulated video, and digital video.

Composite Video: A single video signal that contains luminance (brightness) and chrominance (color) information. The quality of the composite signal is superior to RF modulated video, but is not as good as S-video or component video. The composite video output will most likely be a single RCA-type connector.

Component Video: Three separate video signals are provided on three individual cables of equal length (one brightness Y and two color signals U & V). Component video has increased bandwidth for color information, resulting in a more accurate picture with clearer color reproduction and less bleeding. Analog component video is far superior to composite video and/or S-video and is typically used in production environments.

S-Video: S-video (also known as Y/C) transmits the luminance (brightness) and chrominance (color) information separately. This method eliminates the chroma/luma interference and offers higher quality than *Composite Video*, but lower quality than *Component Video*. It should be noted that the pins on an S-video connector (also known as a 4-pin mini-DIN connector) tend to bend easily and care must be taken when using these connectors.

RF Modulated Video: This signal consists of an RF carrier (Channel 3 or 4) that is modulated with both a composite video signal and an audio signal(s). While *Composite Video, Component Video,* and *S-Video* are all limited by the distance that the signals can travel without amplification, RF modulated video is able to overcome the limitations of distance and can be easily split to provide signals to multiple television sets.

RF modulated video typically travels over coaxial cable that is terminated with F-style connectors, and is demodulated at the television set. RF modulated video provides relatively poor video quality.

This is due to signal loss and degeneration in both the modulation and the demodulation process. However, the inclusion of embedded audio and ease of wiring often outweigh absolute quality as the final decision criteria.

RF modulators are not typically found in satellite receiver products because the RF causes interference with other components in the unit. For that reason, an external RF modulator is often required for locations requiring this method of video distribution.

Digital Video: Digital video outputs are not typically required for business television applications. These outputs are more commonly associated with broadcaster and cable operator applications where the digital video content needs to be routed to other transmission and storage equipment. By keeping this content in a digital state, the broadcaster or cable operator can avoid quality degradation associated with decoding and re-encoding the content.

Products used in broadcast and cable markets may include a serial digital interface (SDI), which is typically transported on coaxial cable using a BNC connector.

The Audio Interfaces for these products consist of the following:

Analog Audio: Playback of analog audio is critical to the success of BTV/IP network. Interface options vary between the products, resulting in differences in audio quality. The products in this guide use XLR, RCA, and stereo miniplugs as their analog audio interfaces.

The XLR interface is shielded and balanced for low noise and low distortion for use in high quality broadcast grade applications. In enterprise applications, XLR connections are often preferred for digital signage applications that includes high impact audio and for large meeting room communications where the audio output of the satellite receiver is often amplified through a local PA system.

The RCA and stereo miniplug connectors provide relatively the same consumer grade quality. The audio signals are unbalanced and tend to be more susceptible to noise. RCA connectors are traditionally common in business television applications where training and corporate communications are displayed to televisions for small to medium sized audiences.

Digital Audio: Much like digital video interfaces, digital audio interfaces are not common in business television. Since the audio is typically intended for playback at the receiver location, as opposed to retransmission to other facilities, most products do not include these interfaces.

5.2.2.2 Television Standards

The manufacturers of BTV/IP media gateways detailed all support the television standards for business television services in the United States. The differences exist in the support of higher quality video and international formats. These differences are due to variations in the market segments targeted by each manufacturer.

The Television Standards section consists of the following five types: NTSC, PAL, PAL-M, PAL-N, and DTV. Additionally, the manufacturers have specified their digital sampling method of 4:2:0 and 4:2:2 within the standards section. The following definitions are intended to clarify the differences between these standards.

BTV/IP MEDIA GATEWAYS > STANDARDS

	ADTEC EDJE-L	HELIUS 2500-S	HELIUS 1500-S	IDC SFX2100	NOVRA SSP 100	SKYSTREAM EVR-7000	WEGENER iPUMP
NTSC	Yes	Yes	Optional	Yes	Yes	Yes	Yes
PAL	Yes	Yes	Optional	Yes	Yes	Yes	Yes
PAL-M	Yes	Optional	Optional	Yes	Yes	Yes	Yes
PAL-N	Yes	Optional	Optional	No	Yes	No	Yes
DTV	No	Optional	Optional	Optional	Optional	No	Optional
4:2:0	Yes	Yes	Optional	Yes	No	Yes	Yes
4:2:2	No	Optional	Optional	Optional	No	No	Optional

Table #7: BTV/IP Media Gateways – Standards

NTSC: An acronym for National Television System Committee, the organization that developed the analog television standard currently in use in the U.S., Canada, and many other nations around the globe. This standard defines a 525-line system with interlaced scan lines at an approximate rate of 30 frames per second.

It is essential that any BTV/IP media gateway in the United States support this standard. All of the products in this guide comply with the standard.

PAL: Phase Alternate Line. This television broadcast standard is common throughout Europe and many other countries. This standard defines a 625-line system with interlaced scan lines at an approximate rate of 25 frames per second.

The ability of a BTV/IP media gateway to support PAL (or any of its derivatives) is important if the network needs to reach nations on the PAL standard. These nations are located throughout the globe. SECAM is also located throughout the world. However, in the majority of these countries, televisions are available that support both PAL and SECAM. Therefore, manufacturers only support PAL.

PAL-M: A derivative of the PAL standard used in most of South America.

PAL-N: A derivative of the PAL standard used in part of South America including Argentina, Paraguay and Uruguay.

DTV (Digital Television): In operation since late 1998 in the United States, this technology allows for the transmission of superb pictures and multi-channel sound as well as embedded data streams. This standard comes in two basic flavors: widescreen, near-film-quality High Definition Television (HDTV), and medium-quality Standard Definition Television (SDTV). HDTV formats range from 1,125-line interlaced (1125i) to 720-line progressive (720p) in 16:9 aspect ratio. SDTV formats include the familiar 525-line interlaced signal in both 4:3 and 16:9 as well as 480-line progressive (480p) in the same aspects.

At the present time, BTV/IP media gateways do not support live HDTV to a television monitor. This is likely to change in the coming years driven by retail networks and their dynamic digital signage applications.

While not considered video standards per se, the digital sampling method does have a dramatic effect on the final video quality. The sampling method defines both how the source signal is encoded and how the local BTV/IP media gateway reproduces the material.

4:2:0: This is the sampling format for the digitization of broadcast quality video and is the basis for MPEG-2 video quality. All of the BTV/IP media gateways in this guide support 4:2:0 video.

Technically, 4:2:0 video digitizes the luminance and color difference components (Y, R-Y, B-Y) of a video signal. The four represents the 13.5 MHz sampling frequency of Y, while the R-Y and B-Y are sampled at 6.75 MHz--effectively between every other line only (one line is sampled at 4:0:0, luminance only, and the next at 4:2:2). This is generally used as a more economical system than 4:2:2 sampling for 625-line formats so that the color signals have a reasonably even resolution in the vertical and horizontal directions for that format.

4:2:2: This is the sampling format for the digitization of component quality video. 4:2:2 video is not typically used in the BTV market. Products that support this tend to be higher end professional grade products. However, some of the products in this guide support this format as an option.

Technically, 4:2:2 video digitizes the luminance and color difference components (Y, R-Y, B-Y) of a video signal. It is generally used as shorthand for ITU-R 601. The term 4:2:2 describes that for every four samples of Y, there are two samples each of R-Y and B-Y, giving more chrominance bandwidth in relation to luminance compared to 4:1:1 sampling. ITU-R 601, 4:2:2 is the standard for digital studio equipment and the terms "4:2:2" and "601" are commonly (but technically incorrectly) used synonymously. The sampling frequency of Y is 13.5 MHz and that of R-Y and B-Y is each 6.75 MHz providing a maximum color bandwidth of 3.37 MHz—enough for high-quality chroma-key. The format specifies eight bits of resolution. Details of the format are specified in the ITU-R BT.601-2 standard document. See also: ITU-R BT.601-2.

5.2.2.3 Storage

	ADTEC EDJE-L	**HELIUS 2500-S**	**HELIUS 1500-S**	**IDC SFX2100**	**NOVRA SSP 100**	**SKYSTREAM EVR-7000**	**WEGENER iPUMP**
BTV/IP MEDIA GATEWAYS > STORAGE							
Hard Drive	40 GB (more opt.)	80 GB (more opt.)	40 GB (more opt.)	40 GB (more opt.)	40 GB standard	80 GB	40 GB (80 GB opt.)
Floppy Drive	No	3.5" 1.44MB	3.5" 1.44MB	No	No	No	No
Flash	No	No	No	No	No	Yes	IDE Flash (opt.)
DVD/CD-ROM	Optional	DVD opt./CD Yes	No	No	Yes	No	Optional

Table #8: BTV/IP Media Gateways – Storage

All BTV/IP media gateway manufacturers provide hard drives for storage. It is this capability that defines these products as a comprehensive single box solution for BTV/IP services. Storage is most critical to video-on-demand applications. The amount of storage in the different products included in this guide varies and is most often scalable as an option.

In addition to a hard drive, this section looks at other drives that are included in some of these products such as floppy drives and DVD/CR-ROM Drives. In addition, flash memory is included.

Hard Drive: Products in this guide offer hard drives that range in size from 40 GB to 240 GB. The only drawback of too much hard drive storage is the additional cost, which is negligible. The amount of video that can be stored on a hard drive is a factor of the length of the video clips and the data rates at which the video is encoded. Table # 4 in Section 3.6 details storage capabilities relative to different hard drive sizes.

If additional storage is required for a BTV/IP service, addition of an external storage device or video file server on the LAN will be required. This is one reason why the amount of storage required for a remote site needs to be determined when deciding the network architecture for a BTV/IP network.

Floppy Drive: Floppy drives offer minimal storage but is inexpensive. Some devices require a floppy drive for maintenance and upgrading the operating system. If one is required your vendor will include one. A floppy drive can be added later, if desired, via USB port.

Flash Memory: Flash memory is expensive but incredibly fast. Many BTV/IP media gateways store their embedded operating systems in flash, causing them to boot quickly, offer some virus protection, and, in the event the hard drive fails, the system will at least boot and give error messages. It is difficult to accidentally erase the operating system and other data from flash, but upgrading the OS is not difficult or time consuming.

DVD/CD ROM: A DVD/CD drive could be used to provide access to program content not available over the network. A drive with write capability can be used to archive program content. If no DVD/CD is built into the system, they are available as add-ons using the gateway's USB interfaces.

5.2.2.4 Computer Interfaces

	ADTEC EDJE-L	HELIUS 2500-S	HELIUS 1500-S	IDC SFX2100	NOVRA SSP 100	SKYSTREAM EVR-7000	WEGENER iPUMP
USB	No	6	6	4	No	2	2
Async Data	RS-232 DE-9P	2 RS-232 DE-9P	2 RS-232 DE-9P*	RS-232 DE-9P	Yes	RS-232 DE-9P	RS-232 DE-9P
Sync Data	X.21 DA-15S	No	No	X.21 DA-15S	No	No	No
VGA Monitor	No	Yes	Yes	Yes	Yes	Yes	Yes
Keyboard	No	PS/2 (opt.)	PS/2 (opt.)	PS/2	No	Via USB	PS/2
Mouse	No	PS/2 (opt.)	PS/2 (opt.)	PS/2	No	No	PS/2
Parallel	Yes	Yes	Yes	1	No	No	No
PCMCIA Slot	No	No	No	No	No	No	No
Terminal	Yes	No	No	RS-232 DE-9P	No	No	Telnet
Modem	No	Yes	Yes	Optional	Yes	Internal (opt.)	Optional
Alarm	No	No	No	Yes	No	No	Relay via DE-9P

BTV/IP MEDIA GATEWAYS > COMPUTER INTERFACES

Table #9: BTV/IP Media Gateways – Computer Interfaces

The BTV/IP media gateway receiver is a special purpose computer system that is a close cousin to a desktop personal computer. As such they all have operating systems, application software and certain input/output ports. Many of these ports are exactly the same as on the personal computer.

There are eleven areas of comparison in this category. They include USB ports, asynchronous data port, synchronous data port, VGA monitor, keyboard, mouse, parallel port, PCMCIA slot, terminal port, internal modem, and alarm port.

USB Port: An acronym for Universal Serial Bus. The USB port was introduced in the late 1990s and allows the computer to support a wide variety of peripheral equipment such as hard drives, DVD/CD ROMs, without overtaxing system resources. Not all operating systems support all USB devices.

The USB port is particularly useful in retail networks. Many deployments of BTV/IP media gateways for digital signage use USB ports to gain data about the in-store environment from monitoring devices such as motion detectors, door alarms, temperature monitors, and pressure sensitive floor mats. Data from these detectors is processed by middleware in the BTV/IP media gateway to determine the most effective video file to be played at a given time.

The more USB ports on a BTV/IP media gateway, the more data points can be used to determine appropriate video files to be displayed.

Asynchronous Data Port: Generally a 9 pin RS-232 standard connector, these ports are standard PC ports that will support modems and other data communication devices attached to them. With a special cable they can attach to other PCs. Theses interfaces do not provide a clock signal and are therefore referred to as asynchronous.

Synchronous Data Port: The Sync (synchronous) data port is an extremely fast, clocked interface not yet found standard on desktop personal computers. They support the X.21 standard. In this application they are used for transferring large video files.

VGA Monitor, Keyboard and Mouse: The monitor, keyboard and mouse ports will function with standard personal computer peripherals.

Parallel Port: Supports a standard personal computer local printer.

PCMCIA Slot: Personal Computer Memory Card Industries Association. More commonly found on laptops than desktop computers there are a wide variety of options that will function in these small option slots, but generally memory cards and modems are the most popular. Some receiver units have optional modems that can be installed in the PCMCIA slot.

However, the BTV/IP Media Gateways use these ports for additional functionality, such as pro-audio outputs or a second video decoder. In general, products with PCMCIA slots have a wider range of available options for customized functionality.

Terminal: Not to be confused with the monitor and keyboard ports, the terminal port attaches only to a "dumb terminal" or a PC running a terminal emulation program such as Hyper-Terminal™. No special cable is required. This is a simple command and control port that generally supports a test only command line interface.

Modem: For the purpose of this Guide, it is being addressed as to whether a receiver supports an internal, external or PCMCIA slot optional modem. All these modems are dial-up and have RJ-11 jacks. These are standard PC modems that support a 56 kbps data rate.

A modem is important to provide a return path to the central hub for deployments that are not connected to a conventional enterprise WAN. The modem is particularly useful in a SOHO deployment.

Alarm: The alarm port is a series of relays that will either open or close when certain conditions exist, such as loss of receive signal. The alarm would then be connected to a local monitoring system.

For most BTV/IP applications, the alarm port is not necessary. Status of the units can be regularly sent back to the Network Operations Center (NOC) using the terrestrial return path. The return path often allows the NOC to be aware of a site problem before the site has detected the problem.

5.2.2.5 Control & Functionality

The requirements for control and functionality with BTV/IP services are far greater than those for traditional BTV. Understanding the control and functionality of a BTV/IP media gateway is critical in determining if any given product is well suited for particular BTV/IP services. These details are often not included in specification sheets. Terminology can be confusing, as different manufacturers use different terms to define the same function.

In this section, the Guide addresses many middleware functions, as well as other control interfaces. Topics in this category include hand held remotes, electronic program guides, on-screen displays, browser interfaces, command line functionality, NMS flexibility, local play configurations, network play configurations, ad insertion, and play logging.

BTV/IP MEDIA GATEWAYS > CONTROL

	ADTEC EDJE-L	HELIUS 2500-S	HELIUS 1500-S	IDC SFX2100	NOVRA SSP 100	SKYSTREAM EVR-7000	WEGENER iPUMP
Hand Held Remote	No	IR	IR	No	No	Yes	RF
EPG w/menu	Yes	Yes	Yes	Yes	Optional	Yes	Yes
OSD VBI Info	Yes	Yes	Optional	No	No	Yes	Yes
OSD Crawl	Yes	Yes	Optional	Yes	Yes	Line by line	Yes
Browser Interface	No	Yes	Yes	Yes	Yes	Yes	Yes
Command Line	Yes	Yes	Yes	Yes	No	Yes	Yes
NMS Flexibility	No	SNMP Based	SNMP Based	SNMP Based	No	SNMP Based	Yes
Local Play Config	Yes	Yes	Optional	Yes	No	Yes	Yes
Network Play Config	Yes	Yes	Optional	Yes	Yes	Yes	Yes
Ad Insertion	Yes	Yes	Optional	Yes	Yes	Yes	Yes
Play Logging	Yes	Yes	Optional	Yes	Yes	Yes	Yes

Table #10: BTV/IP Media Gateways – Control

Hand Held Remote: Hand held remotes, like the one for your TV or VCR at home, provide convenient local control of the receiver. Options depend on the capability of the receiver unit. The most common hand held remote is the infrared type. These send low intensity infrared pulses to a photosensor on the receiver, requiring "line of sight" in order to operate. Other remotes use RF (radio frequency) signals. RF remotes are best for large rooms or other conditions where the receiver is located in a different room than the television. If there is more than one receiver in the area, make sure the RF remotes operate on different frequencies.

One of the major problems with remotes is that they tend to get lost or misplaced at many remote sites. Unfortunately, with the move to BTV/IP this will still be a problem. One solution common with most of the units in this Guide is the ability to access the control functions of the BTV/IP media gateway via a web browser interface from a networked PC.

EPG with Menu: The Electronic Program Guide with Menu provides a simple display on a local TV to select receiver functions and playback control. It is comparable to a cable TV set-top box.

The EPG is an important function for VOD and streaming services. It should be noted that the level of detail about the programming varies between electronic program guides. More sophisticated guides contain advanced metadata about the programming.

OSD VBI Info: This is an acronym for "On Screen Display/Vertical Blanking Interval Information". Within the NTSC and PAL standards there is the capability to transmit text data during the space between frames known as the vertical blanking interval. Much information is transmitted here including closed captioning for the hearing impaired. This option allows the display of whatever content you may choose to place in the VBI for viewing.

67

An important note regarding closed captioning is that government and many corporations are now requiring that this functionality be a part of their networks in order to comply with the Americans with Disabilities Act.

OSD Crawl: On Screen Display Crawl is best compared to the news crawls that appear at the bottom of the screen on cable TV all news and financial channels, along with news, weather and product knowledge information. In the enterprise, this function is used to provide corporate messaging. In retail networks, this can be used for advertising, marketing and promotional information to help enhance the shopper's experience.

Browser Interface: Some BTV/IP media gateways have a browser control interface that can be accessed with either Internet Explorer or Netscape web browsers. This permits desktop remote control of program playback, allowing the user community to download content to the desktop at their convenience.

Command Line: Command line interfaces are powerful tools for the system administrator or other experienced user to program the receiver to perform specialized functions (often associated with digital signage in a BTV/IP environment). Some may require a local control terminal, others may have a Linux operating system allowing a user to log in and open a shell locally or remotely. The command line interface should be accessible only to qualified personnel.

NMS Flexibility: Some receivers will function under any commercially available Network Management System. Others may require proprietary software from the receiver manufacturer. It is very important to fully understand the NMS operations required for BTV/IP services before selecting a technology platform.

Local Play Config: This feature allows for the playing of content directly from the receiver (hard drive) to be controlled at the local site by the local personnel. This may be through a hand held remote, an on screen menu or other local method.

Network managers who do not want to lose centralized control of their networks may not consider this to be a positive feature and elect to not make it available at the receive locations. However, for BTV/IP services to be widely accepted, this feature is often a necessity.

Network Play Config: This feature allows all or some receiver playback capabilities to be controlled from a central network control point. The local user may or may not have or need any control of the receiver playback function.

Ad Insertion: This feature was developed for broadcasters, cable operators and the audio industry and allows advertising to be inserted automatically into program content. Based on the early success of dynamic digital signage networks, it has established a strong place in the enterprise. Retail networks use this function regularly and find that it helps the network generate revenue. This feature can also be used in corporate communications where select "message ads" are regularly inserted in the programming of employee information channels.

Play Logging: Play logging creates a record of what programs and advertising were played when and on what displays. This can be used to verify contractual obligations and create billing records. It can also be used to help determine the effectiveness of video messaging and assist organizations in maximizing this effectiveness.

5.2.2.6 Operating Platform

Historically, the operating system has been a limiting factor in the development of BTV/IP services. This was very apparent to anyone who attempted to use a printed circuit board style satellite receiver for personal computers such as those manufactured in the mid-to-late 1990s. Networks based on these technologies found the operating systems of the time (such as Windows 95) to be unreliable and the PCs regularly required rebooting or having their operating systems reinstalled.

Today, this has been overcome by newer, more reliable, operating systems. This section discusses the operating systems for BTV/IP media gateways as well as three other parameters: power, form factor, and memory.

BTV/IP MEDIA GATEWAYS > PLATFORM

	ADTEC EDJE-L	HELIUS 2500-S	HELIUS 1500-S	IDC SFX2100	NOVRA SSP 100	SKYSTREAM EVR-7000	WEGENER iPUMP
Operating System	VTOS (Proprietary)	Linux	Linux	Linux	Embedded XP	Linux	Linux
Power	115 VAC 60 Hz 230 VAC 50 Hz	100-125 or 200-240 VAC 50-60 Hz	100-125 or 200-240 VAC 50-60 Hz	Universal AC Power, Autosensing	110-220 auto	100-125 or 200-240 VAC 50-60 Hz 80 Watts	115 VAC 60 Hz 230 VAC 50 Hz
Form Factor	1 RU	Desktop	Desktop	1 RU or Desktop	2U rack	Desktop	2 RU
Memory (RAM)	16 MB	256 MB	256 MB	256 MB (more opt.)	As required	128 MB	128 MB (more opt.)

Table #11: BTV/IP Media Gateways – Platform

Operating System: The IP Receiver Gateway or Media Router is essentially a high-powered purpose built computer, and as such needs an operating system to interface between the application software and the hardware. The operating system most users are familiar with is Microsoft Windows. But there are many others, such as UNIX, Linux, and VX-Works. Some manufacturers may choose to create their own operating systems.

Linux is the emerging favorite for high-powered network and server applications. Very much like UNIX, Linux provides a small, fast, inexpensive and reliable system kernel that has a growing following in the IT field. The Linux operating system has been the catalyst for the development of "deployable" BTV/IP products and services

Microsoft has countered Linux with Windows XP, which is being used in one of the products included in this BTV/IP media gateway comparison.

Power: In the U.S. and Canada most systems require 110 – 120 VAC power. For export to Europe most systems provide a means to adapt to 220 – 240 VAC by means of a small switch on the real panel of the equipment case.

Some systems require a small DC voltage, between 12 and 24 VDC. In this case, the manufacturer will provide a wall transformer or other external power supply. For Europe, a different external power supply will be provided.

Form Factor: For the purpose of this Guide, it means 'shape'. Does the equipment sit on a desk or does it fit into a standard 19-inch rack? And if it fits in a rack, how many 1.75-inch high 'rack units'

does it require? Different form factors are preferred by different enterprises depending on the remote environment.

Memory (RAM): RAM is an acronym for Random Access Memory. Generally speaking, the more RAM a system has available to its CPU the better the system will perform when running complex concurrent applications. The only drawback of too much RAM is incremental cost.

It should be noted that some operating systems require more RAM than others, which should be factored into the decision-making process.

5.2.2.7 TCP/IP Support

Any BTV/IP Satellite Receiver sits on a local area network and must function as a network appliance using the TCP/IP protocols. In addition, the receiver must have router functionality in order to distribute the content coming off the satellite onto the LAN. This section compares seven TCP/IP related categories relevant to these products.

Transmission Control Protocol/Internet Protocol is the worldwide standard Internet protocol suite developed by the U.S. Department of Defense in the 1970s. TCP governs the exchange of sequential data. IP routes outgoing and recognizes incoming messages. This is the protocol stack that makes the Internet and most high-speed computer networks possible. There are many sub-protocols in the TCP/IP network model, such as, Simple Network Management Protocol (SNMP), and Hyper Text Transfer Protocol (HTTP), which is the protocol your web browser tool uses to manage and display files from the Internet.

BTV/IP MEDIA GATEWAYS > TCP/IP

	ADTEC EDJE-L	HELIUS 2500-S	HELIUS 1500-S	IDC SFX2100	NOVRA SSP 100	SKYSTREAM EVR-7000	WEGENER iPUMP
RJ-45 Ports	1	1, 2nd optional	1, 2nd optional	2	1	1	1, 2nd optional
IP Streaming	Yes	Yes	Yes	Yes	Yes	Yes	Yes
TCP/IP Acceleration	No	Yes	Yes	Optional	No	No	Yes
Multicast	Yes	Yes	Yes	Yes	Yes	Yes	Yes
IGMP Grouping	Yes	Yes	Yes	Yes	No	Yes	Yes
SNMP	No	Yes	Yes	Yes	Yes	Yes	Yes
Data Rate	9 Mbs	up to 80 Mbs	up to 80 Mbs	70 Mbs	40 Mbps	up to 72.5 Mbs	up to 80 Mbs

Table #12: BTV/IP Media Gateways – TCP/IP

RJ-45 Port: Short for Registered Jack-45, it is an eight-wire connector used commonly to connect computers onto LANs, especially Ethernets. RJ-45 connectors look similar to the ubiquitous four-wire RJ-11 connectors used for connecting telephone equipment, but they are somewhat wider. If a device has two or more RJ-45 ports, there is a corresponding electrical interface and IP address for each port.

Multicast traffic and streaming of large video files often proves problematic for many enterprises because these applications take up bandwidth impacting the quality of service of other applications running on the LAN. As a result, some organizations run a separate LAN for multicast and video traffic.

A receiver with multiple RJ-45 ports can integrate seamlessly onto these networks, allowing users to browse program guides and search for files on the primary LAN while watching streaming video on the multicast LAN.

IP Streaming: Multimedia content such as video, audio, text, or animation, that is received from the Internet, broadcast network, or local storage device and is distributed to local users via a TCP/IP LAN. All of the products in this Guide support IP streaming.

TCP/IP Acceleration: This term is an acronym for Transmission Control Protocol/Internet Protocol (TCP/IP) Acceleration. This is the controller that is inserted between the Memory Access Layer (MAL) and an Ethernet Media Access Controller (EMAC). The Acceleration controller provides hardware acceleration functions for TCP/IP to improve bandwidth and lower processor core utilization by offloading many TCP/IP protocol stack functions away from the Central Processing Unit (CPU).

This allows for better usage of CPU time and faster communication with the Network Interface. These functions include checksum verification for TCP/User Datagram Protocol (UDP)/IP headers in the receive path, checksum generation for TCP/UDP/IP headers and TCP segmentation support in the transmit path.

This is an important feature to IT departments when considering if video traffic can be supported on the LAN.

Multicast: Data flow from a single source to multiple destinations in a manner very similar to a television broadcast. However, a multicast may be distinguished from a broadcast in that the number of destinations may be limited. Not to be confused with a term often used incorrectly to describe digital television program multiplexing. All of the products in this guide support multicasting.

IGMP Grouping: Internet Group Management Protocol is defined in RFC 1112 as the standard for IP multicasting in the Internet. It's used to establish host memberships in particular multicast groups on a single network. The mechanisms of the protocol allow a host to inform its local router, using Host Membership Reports, that it wants to receive messages addressed to a specific multicast group. All hosts conforming to level 2 of the IP multicasting specification require IGMP. This is an important feature to many IT departments when considering if video traffic can be supported on the LAN.

5.2.2.8 Security

Security issues in a BTV/IP environment are more complex than they are in a traditional BTV environment. BTV networks provide encryption and conditional access at the DVB layer, protecting the entire satellite carrier from the outside, but not the individual programs once they are de-encrypted and de-scrambled by the satellite receiver.

The users of BTV networks also knew that the receiver was part of the same technology platform as the DVB equipment at the network's head end. As a result, the conditional access system could be imbedded in the receiver negating the need for peripheral security items such as smart cards.

71

In today's BTV/IP environment security extends beyond the DVB layer to the individual IP streams contained within a DVB carrier. In addition, there are emerging new security standards such as MPEG-21 that will provide digital rights management to the video objects contained within the IP streams. Lastly, as the BTV/IP satellite receivers reside on both local and wide area networks, their content contribution may come from multiple sources that are not part of the same technology platform. This creates the need for security devices such as smart cards.

BTV/IP MEDIA GATEWAYS > SECURITY

	ADTEC EDJE-L	HELIUS 2500-S	HELIUS 1500-S	IDC SFX2100	NOVRA SSP 100	SKYSTREAM EVR-7000	WEGENER iPUMP
IP Security	Proprietary	CAS	CAS	CAS	No	No	CAS
DVB Layer	CAS	CI CAS	CI CAS	CI CAS	Yes	No	CI CAS
CA DRM	No	No	No	No	No	No	Yes
Smart Card	No	Yes	Yes	Yes	No	Yes	Yes

Table #13: BTV/IP Media Gateways – Security

IP Security: Conditional access security (CA or CAS) is a technology used to control access to digital television services to authorized users by encrypting the transmitted programming. CA has been used for years for pay-TV services. CA is not designed solely for direct-to-home (DTH) broadcasting. It can be used for digital radio broadcasts, digital data broadcasts, non-broadcast information, and interactive services.

A typical CA process involves three basic elements: the broadcast equipment, the set-top box, and the security module. The broadcast equipment generates the encrypted programs that are transmitted to the authorized viewer. When these are transmitted, the set-top box filters out the signals and passes them to the security module. The security module then authorizes these programs for decryption. The programs are then decrypted in real time and sent back to the set-top box for display.

IPsec or IP Security. The protocol used to give authentication and/or encryption to IP packets. IPsec provides security services at the IP layer by enabling a system to select required security protocols, determine the algorithm(s) to use for the service(s), and put in place any cryptographic keys required to provide the requested services. IPsec can be used to protect one or more "paths" between a pair of hosts, between a pair of security gateways, or between a security gateway and a host.

The set of security services that IPsec can provide includes access control, connectionless integrity, data origin authentication, rejection of replayed packets (a form of partial sequence integrity), confidentiality (encryption), and limited traffic flow confidentiality. Because these services are provided at the IP layer, they can be used by any higher layer protocol, e.g., TCP, UDP, ICMP, BGP, etc.

IPsec is ideal for use in virtual private networks (VPN) or tunneling environments.

DVB Layer: Conditional access is not wholly specified in DVB, but a series of tools enable users of DVB to come up with the most effective and efficient mechanisms for their market. Key to DVB CA is the DVB common scrambling algorithm.

In the DVB layer security, there are two CA scenarios: Simulcrypt and Multicrypt.

SimulCrypt is a mechanism whereby a single transport stream can contain several CA systems. This enables different CA decoder populations (potentially with different CA systems installed) to receive and correctly decode the same video and audio streams.

MultiCrypt revolves around the specification of a common interface which when installed in the set-top-box or television permits the user to switch manually between CA systems.

CA DRM: Conditional Access, Digital Rights Management. DRM solutions are software-based platforms that facilitate ease of file management, performing important content security features:

1. Provide Conditional Access to Content: Using DRM, access to content can be limited by a variety of conditions.

2. Guarantee Copyright Protection: With file-level encryption, DRM ensures that malicious users/competitors are unable to modify or alter content that is specifically covered under legal copyright protections.

3. DRM allows you to sell your streaming media using a Pay-to-Play business model that prevents anyone from accessing the file without first having to remit payment, or have a billing arrangement, regardless of how they receive the file. Streaming media/web casts, web site downloads, CD/DVD distribution, or peer to peer file sharing are all controlled, including the number of times the file can be accessed.

There are a wide range of DRM solutions available. It is important to explore the specifics of your unique application.

Smart Card: Conforming to ISO 7816 standards the Smart Card is a portable programmable device containing an integrated circuit, which stores and processes information. Smart Cards look like credit cards or one of those access key cards used to open security doors. A Smart Card contains an encrypted key that can control access to a network, its management system and control how data is encrypted. If Smart Card security is required, then a receiver that has a card reader built option will be a requirement. The presence of a reader does not mean the feature must be used, as it can be disabled.

There are drawbacks to Smart Card security. A card could be lost or stolen. Some personnel will leave the card in the reader at all times, thus defeating the purpose of the system and exposing the card to theft. A Smart Card cannot be copied.

5.2.2.9 Receiver Specifications

The satellite receiver specifications need to be considered during the decision making process. Satellite receivers have earned a reputation for being highly reliable. Although this is true, not all satellite receivers are the same and this section will provide value when selecting a technology platform.

Some manufacturers tout the high data throughput of their receivers. The question really is how small of a carrier can the receiver successfully downlink. Other considerations are the types of modulation schemes that can be supported and how does the receiver react when the signal is lost. This section explores these issues.

	ADTEC EDJE-L	HELIUS 2500-S	HELIUS 1500-S	IDC SFX2100	NOVRA SSP 100	SKYSTREAM EVR-7000	WEGENER IPUMP
RF Input	F-type, 75 ohm	F-type, 75 ohm	F-type, 75 ohm	F-type, 75 ohm	F-type, 75 ohm	F-type, 75 ohm	F-type, 75 ohm
Input Level	-25 to 65 dBm	-25 to -65 dBm	-25 to -65 dBm	-35 to -65 dBm	-25 to -65 dBm	-25 to -65 dBm	-25 to 65 dBm
Freq. Range	950 - 2150 mHz	950 - 2150 mHz	950 - 2150 mHz	950 - 2150 mHz	950 - 2150 mHz	950 - 2150 mHz	950 - 2150 mHz
LNB Power	13 - 18 VDC	13 -18 VDC	13 -18 VDC	13 - 18 VDC	13 -18 VDC	13 - 18 VDC	13 - 18 VDC
LNB Control	22 kHz	22 kHz	22 kHz	22 kHz	22 kHz	22, 33, 44 kHz	22 kHz
FEC	Virterbi 1/2, 2/3, 3/4, 5/6, 7/8 — Reed-Solomon (204, 188)	Virterbi 1/2, 2/3, 3/4, 5/6, 7/8	Virterbi 1/2, 2/3, 3/4, 5/6, 7/8	Virterbi 1/2, 2/3, 3/4, 5/6, 7/8 — Reed-Solomon (204, 188)	Virterbi 1/2, 2/3, 3/4, 5/6, 7/8 (autosense) Reed-Solomon (204, 188)	Virterbi 1/2, 2/3, 3/4, 5/6, 7/8 (autosense) Reed-Solomon (204, 188) T=8	Virterbi 1/2, 2/3, 3/4, 5/6, 7/8 — Reed-Solomon (204, 188)
Demodulator	QPSK, BPSK	QPSK	QPSK	BPSK, QPSK, (16QAM, 8PSK Optional)	QPSK	QPSK	QPSK
Symbol Rate	2 -45 Msps	2 - 50 Msps	2 - 50 Msps	256Ksps - 45Msps	1.5 - 45 Msps	2 - 45 Msps	1.55 - 45 Msps

Table #14: BTV/IP Media Gateways – Receiver Specifications

RF Input: All of the receivers in this comparison have a standard 75-ohm F connector that accepts 75-ohm cable television RG-59 or equivalent coaxial cable.

Input Level: The desired signal level varies only slightly between units. Receive signal level, measured at the back of the unit, is critical to error free operation of the receiver.

Frequency Range: All of the receivers in this comparison are L-band units. The LNB will translate the satellite downlink frequency to somewhere in the range of 950 – 2150 MHz. The proper LNB must be selected depending on the satellite in use: C-band, Ku-band, etc.

LNB Power: The Low Noise Block (down converter and amplifier package located on the antenna) gets its power from the receiver device. Generally the receiver will offer two voltage selections and "OFF." If more than one receiver shares the LNB, only one will provide power. The DC voltage is placed on the center conductor of the coaxial cable. Be aware of the presence of DC voltage on the cable so as not to short it out and possibly damage the receiver or LNB. Always turn off power to the receiver device before disconnecting or reconnecting the coaxial cable. The LNB must be compatible with the receiver.

LNB Control: This is a low frequency signal that is routed from the receiver to the input of an LNB. Without the need for a separate control line, this signal can be used, for example, to switch the frequency range of a universal LNB or, with the help of a switch box, to switch between two different LNBs. Most receivers and LNBs use 22 kHz but other frequencies are available. The receiver must be compatible with the LNB.

FEC: Forward Error Correction refers to a group of techniques for controlling errors in a one-way communication system, such as a satellite system. FEC sends extra information along with the data, which can be used by the receiver to check and correct the data. FEC can be one of many complex algorithms; the most popular are Reed-Solomon and Virterbi. The FEC ratio can be adjusted to compensate for noise on the satellite link.

Demodulator: For the purpose of this Guide, the interest in how many modulation schemes the receiver can demodulate, and not comparing the relative merits of each. To select a receiver that can demodulate the carrier generated by the uplink equipment.

>**QPSK:** Quadrature phase shift keying. QPSK is a digital frequency modulation technique used for sending data over coaxial cable networks. Since it's both easy to implement and fairly resistant to noise, QPSK is used primarily for sending data from the cable subscriber upstream to the Internet.

>**BPSK:** Biphase shift keying. BPSK is a digital frequency modulation technique used for sending data over a coaxial cable network. This type of modulation is less efficient--but also less susceptible to noise--than similar modulation techniques, such as QPSK and 64QAM.

>**16QAM:** A more advanced type of Quadrature amplitude modulation. A complex but highly efficient downstream digital modulation technique that conforms to the International Telecommunications Union (ITU) standard ITU-T J. 83 which calls for quadrature amplitude modulation (QAM) with concatenated trellis coded modulation, plus enhancements such as variable interleaving depth for low latency in delay sensitive applications such as data and voice.

>**8PSK:** yields more throughput than QPSK by doubling the amount of data that can be encoded on the satellite carrier using the same amount of bandwidth.

Symbol Rate: The channel width is just another way of saying symbol rate. Do not confuse with bit rate, as one symbol can translate into many bits. In QPSK modulation, for example, both the phase and amplitude of the carrier wave are changed (modulated) so that one set of changes in the carrier, called a symbol, can represent the encoding of many bits. Symbol rate is not to be confused with throughput on the TCP/IP network. The two are unrelated.

5.3   IP Satellite Routers

An IP satellite router is a one-way satellite receiver that provides an IP network interface with router-type functionality capable of providing IP video streaming onto an enterprise local area network. These products have neither the audio/video outputs nor storage capabilities of a BTV/IP media gateway. These products provide more robust routing and have higher throughputs than IP satellite receivers.

5.3.1   IP Satellite Router Product Overviews

The IP satellite router products included in this guide are the International Datacasting SRA2100, IPricot S1100, IPricot S1000, and SkyStream EMR-5500.

5.3.1.1 International Datacasting SRA2100

The SRA2100 is a fully functional IP satellite router manufactured by International Datacasting. This product differentiates itself from its competition in many key areas, most notably the quality of its components and design.

Key components of the SRA2100 has a superior satellite receiver capable of receiving 8PSK & 16QAM modulated signals, and is able to receive carriers at 256 kbps – lower than any other product in this guide. This allows customers with relatively small datacasting requirements to operate at lower satellite bandwidths, thereby reducing network costs.

International Datacasting also provides its own content management system, Datacast XD. This system allows great flexibility in conditional access and forward error correction schemes. It also allows control over return channels, event scheduling, and stream management.

5.3.1.2   IPricot S1100

The IPricot S1100 is an IP satellite router offering high-speed data connectivity for applications such as streaming audio and video, IP multicast file transfer, fast internet, and data push services such as stock market feeds.

The S1100 features include flash memory and support of QPSK, 8PSK, and 16QAM. This product design focus is its ability to download large amounts of data at very high speeds. This makes it ideal for terrestrial WAN overlay deployments.

5.3.1.3   IPricot S1000

The IPricot S-1000 is an IP satellite router well suited for many BTV/IP applications. Formerly called the DotLink-S, this product has been on the market since 1996. It is deployed globally in large volumes and enjoys a reputation as a reliable and complete solution.

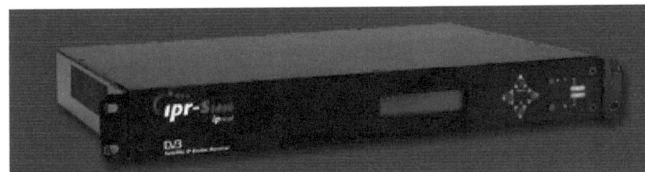

The IPricot S-1000 differs from the IPricot S-1100 in two areas: the ability to download at very high speeds (the S-1000 can downlink at 72 MBPS while the S-1100 can downlink at up to 100 MBPS) and the ability to receive only QPSK modulated signals. These differences are not significant to most BTV/IP services.

5.3.1.4 SkyStream EMR-5500

The SkyStream EMR-5500 is an edge media router with a reputation for reliability and quality performance. This product is currently deployed in many BTV/IP networks.

Residing at the edge of enterprise networks or in service provider points-of-presence (POPs), the EMR-5500 extracts IP content from incoming DVB MPEG-2 transports streams. It routes content to last-mile broadband networks for delivery over 10/100 Fast Ethernet connections or other WAN interfaces. Supporting a wide range of intelligent distribution capabilities, network topologies and performance options, the EMR-5500 is fully integrated with SkyStream 's Source Media Routers, zBand® Content Delivery Platform and E-Manager management suite.

5.3.2   IP Satellite Router Product Comparisons

This section of the guide provides comparisons of the IP satellite routers included in this report. In order to best accomplish this, we compare the products in six categories, including:

- Computer Interfaces
- Control & Functionality
- Operating Platform
- TCP/IP Support
- Security
- Receiver Specifications

In each of these categories, we provide a table comparing the products against key parameters in that category. We then provide notes on the products discussing key points relevant to specific category. Definitions of the particular parameters are references back to Section 5.2.2.

5.3.2.1   Computer Interfaces

IP SATELLITE ROUTERS > COMPUTER INTERFACES

	IDC SRA2100	IPRICOT S1100	IPRICOT S1000	SKYSTREAM EMR-5500
USB	4 (opt.)	No	No	No
Async Data	RS-232 DE-9P	RS-232 DE-9P	RS-232 DE-9P	RS-232 DE-9P
Sync Data	X.21 DA-15S	No	No	Option PCI Slot
VGA Monitor	Optional	No	No	No
Keyboard	Optional	No	No	No
Mouse	Optional	No	No	No
Parallel	Optional	1	1	No
PCMCIA Slot	No	No	No	2
Terminal	RS-232 DE-9P	RS-232 DE-9P	RS-232 DE-9P	Yes
Modem	Optional	Optional	Optional	PCI Slot Option
Alarm	Yes	No	No	No

Table #15: IP Satellite Routers – Computer Interfaces

The IP satellite router is a special purpose computer system that is a close cousin to a desktop personal computer. As such they all have operating systems, application software and certain input/output ports. Many of these ports are exactly the same as on the personal computer. The definitions for these products are detailed in Section 5.2.2.4.

There are eleven areas of comparison in this category. They include USB ports, asynchronous data port, synchronous data port, VGA monitor, keyboard, mouse, parallel port, PCMCIA port, terminal port, internal modem, and alarm port.

5.3.2.2   Control & Functionality

The requirements for control and functionality with BTV/IP services are far greater than those for traditional BTV. Understanding the control and functionality of an IP satellite router is critical in determining if any particular product is well suited for particular BTV/IP services. These details are often not included in specification sheets. Terminology is often confusing, as different manufacturers use different terms to define the same function.

Included in this section are many middleware functions, as well as other control interfaces. Topics in this category include hand held remotes, electronic program guides, on-screen displays, browser interfaces, command line functionality, NMS flexibility, local play configurations, network play configurations, ad insertion, and play logging. Some of these functions are not possible with an IP satellite router, but are included to show how these products differ from BTV/IP media gateways.

IP SATELLITE ROUTERS > CONTROL

	IDC SRA2100	IPRICOT S1100	IPRICOT S1000	SKYSTREAM EMR-5500
Hand Held Remote	No	No	No	No
Browser Interface	Yes	Yes	Yes	Yes
Command Line	Yes	Yes	Yes	Yes
NMS Flexibility	SNMP Based	SNMP Based	SNMP Based	SNMP Based
Local Play Config	Yes	Yes	Yes	Yes
Network Play Config	Yes	Yes	Yes	Yes
Play Logging	No	Yes	Yes	Yes

Table #16: IP Satellite Routers – Control

5.3.2.3   Operating Platform

Historically, the operating system has been a limiting factor in the development of BTV/IP services. This was very apparent to anyone who attempted to use a printed circuit board style satellite receiver for personal computers such as those manufactured in the mid-to-late 1990's. Networks based on these technologies found the operating systems of the time (such as Windows 95) to be unreliable and the PC's regularly required rebooting or having their operating systems reinstalled.

Today, this has been overcome by newer, more reliable, operating systems. This section discusses the operating systems for IP satellite routers as well as three other parameters: power, form factor, and memory. Definitions are included in Section 5.2.2.6

IP SATELLITE ROUTERS > PLATFORM

	IDC SRA2100	IPRICOT S1100	IPRICOT S1000	SKYSTREAM EMR-5500
Operating System	Linux	Net BSD	Net BSD	Linux
Power	Universal AC Power, Autosensing	83 to 264 VAC 47 to 63 Hz, 60 Watts	83 to 264 VAC 47 to 63 Hz, 60 Watts	100-125 or 200-240 VAC 50-60 Hz, 80 Watts
Form Factor	1 RU or Desktop	1 RU	1 RU	1 RU

Table #17: IP Satellite Routers – Platform

5.3.2.4   TCP/IP Support

Any BTV/IP Satellite Receiver sits on a local area network and must function as a network appliance using the TCP/IP protocols. In addition, the receiver must have router functionality in order to distribute the content coming off the satellite onto the LAN. This section compares seven TCP/IP related categories relevant to these products.

All of the comparison categories in this section are of importance in selecting an IP satellite router. Definitions of the comparison categories are included in Section 5.2.2.7.

IP SATELLITE ROUTERS > TCP/IP				
	IDC SRA2100	IPRICOT S1100	IPRICOT S1000	SKYSTREAM EMR-5500
RJ-45 Ports	2	1	1	2
IP Streaming	Yes	Yes	Yes	Yes
TCP/IP Acceleration	Optional	Yes	Yes	Optional
Multicast	Yes	Yes	Yes	Yes
IGMP Grouping	Yes	Yes	Yes	Yes
SNMP	Yes	Yes	Yes	Yes
Throughput	70 Mbs	100 Mbs	72 Mbs	72.5 Mbs

Table #18: IP Satellite Routers – TCP/IP

5.3.2.5   Security

Security issues in a BTV/IP environment are more complex than they are in a traditional BTV environment. Traditional BTV networks viewed encryption and conditional access at the DVB layer, protecting the entire satellite carrier from the outside, but not the individual programs once they are de-encrypted and de-scrambled by the satellite receiver.

Traditional BTV networks also knew that the receiver was part of the same technology platform as the DVB equipment at the network's head end. As a result, the conditional access system could be imbedded in the receiver negating the need for peripheral security items such as smart cards.

Definitions for the comparison categories are included in Section 5.2.2.8.

IP SATELLITE ROUTERS > SECURITY				
	IDC SRA2100	IPRICOT S1100	IPRICOT S1000	SKYSTREAM EMR-5500
IP Security	CAS	IPCrypt, IPSec	IPCrypt, IPSec	IPSec
DVB Layer	CI CAS	Optional	Optional	No
CA DRM	No	Optional	Optional	Optional
Smart Card	Optional	Optional	Optional	Optional

Table #19: IP Satellite Routers – Security

5.3.2.6   Receiver Specifications

The satellite receiver specifications need to be considered during the decision making process. Satellite receivers have earned a reputation for being highly reliable. Although this is true, not all satellite receivers are the same and this section will provide value when selecting a technology platform.

Comparison Category Definitions are included in Section 5.2.2.9.

IP SATELLITE ROUTERS > RECEIVER SPECIFICATIONS

	IDC SRA2100	IPRICOT S1100	IPRICOT S1000	SKYSTREAM EMR-5500
RF Input	F-type, 75 ohm	F-type, 75 ohm	F-type, 75 ohm	F-type, 75 ohm
Input Level	-35 to -65 dBm	-25 to -65 dBm	-25 to -65 dBm	-25 to -65 dBm
Freq. Range	950 - 2150 mHz	950 - 2150 mHz	950 - 2150 mHz	950 - 2150 mHz
LNB Power	13 - 18 VDC	14 -18 VDC	14 -18 VDC	13 - 18 VDC
LNB Control	22 kHz	22 kHz	22 kHz	22, 33, 44 kHz
FEC	Virterbi 1/2, 2/3, 3/4, 5/6, 7/8 Reed-Solomon 204, 188	Virterbi 1/2, 2/3, 3/4, 5/6, 7/8 Reed-Solomon 204, 188	Virterbi 1/2, 2/3, 3/4, 5/6, 7/8 Reed-Solomon 204, 188	Virterbi 1/2, 2/3, 3/4, 5/6, 7/8 Reed-Solomon 204, 188
Demodulator	BPSK, QPSK (8PSK, 16QAM Optional)	QPSK, 8PSK, 16QAM	QPSK	QPSK, BPSK
Symbol Rate	256Ksps - 45Msps	1 -45 Msps	1 -45 Msps	2 - 45 Msps

Table #20: IP Satellite Routers – Receiver Specifications

5.4   IP Satellite Receivers

An IP satellite receiver is a one-way satellite receiver that provides an IP network interface with router-type functionality capable of providing IP video streaming to a SOHO local area network. These products have neither the audio / video outputs or storage capabilities of a BTV/IP media gateway. These products have less throughput than IP satellite routers making then more suited for SOHO deployments than in a large enterprise.

5.4.1   IP Satellite Receiver Product Overviews

The IP satellite router products included in this guide are the International Datacasting SRA2000plus, Ipricot Sc, Ipricot S500+, Mainstream Data DVB+, Novra S75, and Skystream EMR-1600.

5.4.1.1 International Datacasting SRA2000plus

The SRA2000plus is an IP satellite receiver manufactured by International Datacasting. This product differentiates itself from its competition in many key areas, most notably the quality of its components and design.

Key components of the SRA2000plus has a superior satellite receiver capable of receiving 8PSK & 16QAM modulated signals, and is able to receive carriers at 256 kbps – lower than any other product in this guide. This allows customers with relatively small datacasting requirements that are of great significance to enterprises reaching SOHO environments.

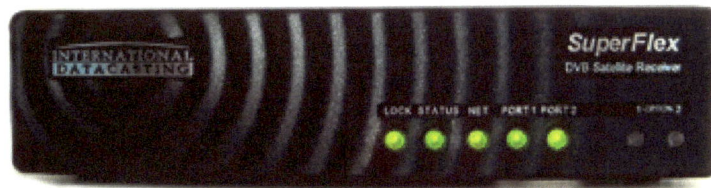

International Datacasting also provides its own content management system, Datacast XD. This system allows great flexibility in conditional access and forward error correction schemes. It also allows control over return channels, event scheduling, and stream management.

5.4.1.2 Ipricot Sc

The Ipricot Sc is an IP satellite receiver designed to provide SOHO and digital signage markets with content distribution and terrestrial WAN overlay applications. The Ipricot Sc is designed to be a low cost solution with some high level features.

The Ipricot Sc features include flash memory and support throughputs up to 8 MBPS. This product is reliable and well suited for smaller type applications such as large file transfer and multimedia streaming.

5.4.1.3 Ipricot S500

The Ipricot S500 is an IP satellite receiver designed to provide small enterprise, SOHO and digital signage markets with LAN-based BTV, content distribution and terrestrial WAN overlay applications. The Ipricot S500 is designed to be a mid-range solution that for networks with requirements that range somewhere between Ipricot's Sc and S1000 products

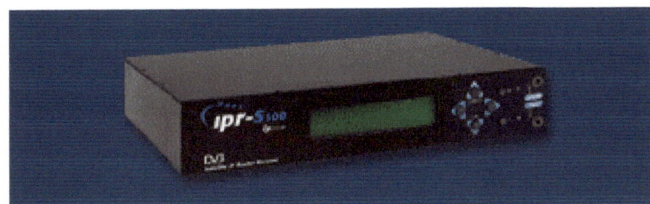

The Ipricot S500 features include flash memory and support throughputs up to 40 MBPS. This product is reliable and further differentiates itself from the Ipricot Sc by having more sophisticated LAN management/ routing capabilities.

5.4.1.4   Mainstream Data DVB+

The Mainstream DVB+ is an IP satellite receiver intended for the SOHO and small enterprise environments. This product differentiates itself with its network management capabilities.

Mainstream Data has been in the satellite receiver market since 1985. Mainstream was a pioneer in FM radio data broadcasting. Their products are noted for their reliability and are deployed extensively throughout North America, Europe, and Asia.

5.4.1.5   Novra S75

The Novra S75 is an IP satellite receiver designed to provide small enterprise, SOHO and digital signage markets. The product is targeted at applications such as streaming video/multimedia, high-speed internet downloading, and file distribution.

Novra S75 features include powerful download throughput of 55 MBPS and the ability to have the operating system of your choice loaded onto it at the time of purchase. This product is a small market extension of the Novra product line that includes two BTV/IP media gateways contained in this Guide.

5.4.1.6 Skystream EMR-1600

The Skystream EMR-1600 is an IP satellite receiver with some media router capabilities targeted at the SOHO market with up to 16 users.

A clear differentiator for this product is the EMR-1600's full integration with SkyStream's Source Media Routers, zBand® Content Delivery Platform and E-Manager management suite. This provides network with great management flexibility.

5.4.2 IP Satellite Receiver Product Comparisons

This section of the guide provides comparisons of the IP satellite receivers included in this report. In order to best accomplish this, we compare the products in six categories, including:

- Computer Interfaces
- Control & Functionality
- Operating Platform
- TCP/IP Support
- Security
- Receiver Specifications

In each of these categories, we provide a table comparing the products against key parameters in that category. We then provide notes on the products discussing key points relevant to specific category. Definitions of the particular parameters are references back to Section 5.2.2.4

5.4.2.1   Computer Interfaces

IP SATELLITE RECEIVERS > COMPUTER INTERFACES

	IDC SRA2000	IPRICOT SC+	IPRICOT S500+	MAINSTREAM DVB+	NOVRA S75	SKYSTREAM 1600
USB	No	No	No	No	No	No
Async Data	RS-232 DE-9P	RS-232 DE-9P	RS-232 DE-9P	RS-232 DE-9P	No	RS-232 DE-9P
Sync Data	No	No	No	No	No	No
VGA Monitor	No	No	No	No	No	No
Keyboard	No	No	No	No	No	No
Mouse	No	No	No	No	No	No
Parallel	No	No	No	No	No	No
PCMCIA Slot	No	No	No	No	No	No
Terminal	Yes	RS-232 DE-9P	RS-232 DE-9P	No	No	No
Modem	No	No	No	No	No	No
Alarm	No	No	No	No	No	No

Table #21: IP Satellite Receivers – Computer Interfaces

The IP satellite receiver is a special purpose computer system that is a close cousin to a desktop personal computer. As such they all have operating systems, application software and certain input/output ports. Many of these ports are exactly the same as on the personal computer. The definitions for these products are detailed in Section 5.2.2.4.

There are eleven areas of comparison in this category. They include USB ports, Asynchronous data port, synchronous data port, VGA Monitor, Keyboard, Mouse, Parallel port, PCMCIA Port, Terminal port, internal modem, and alarm port.

5.4.2.2   Control & Functionality

The requirements for control and functionality with BTV/IP services are far greater than those for traditional BTV. Understanding the control and functionality of an IP satellite router is critical in determining if any particular product is well suited for particular BTV/IP services. These details are often not included in specification sheets.   Terminology is often confusing, as different manufacturers use different terms to define the same function.

Included in this section are many middleware functions, as well as other control interfaces. Topics in this category include hand held remotes, electronic program guides, on-screen displays, browser interfaces, command line functionality, NMS flexibility, local play configurations, network play configurations, ad insertion, and play logging. Some of these functions are not possible with an IP satellite router, but are included to show how these products differ from BTV/IP media gateways.

IP SATELLITE RECEIVERS > CONTROL

	IDC SRA2000	IPRICOT SC+	IPRICOT S500+	MAINSTREAM DVB+	NOVRA S75	SKYSTREAM 1600
Hand Held Remote	No	No	No	No	No	No
EPG w/menu	No	No	No	No	No	No
OSD VBI Info	No	No	No	No	No	No
OSD Crawl	Yes	No	No	No	No	No
Browser Interface	Yes	Yes	Yes	Yes	Yes	Yes
Command Line	Yes	Yes	Yes	No	No	No
NMS Flexibility	SNMP Based	SNMP Based	SNMP Based	No	SNMP Based	SNMP Based
Local Play Config	Yes	Yes	Yes	Yes	via LAN	Yes
Network Play Config	Yes	Yes	Yes	Yes	Yes	No
Ad Insertion	No	No	No	No	No	No
Play Logging	No	No	No	No	No	No

Table #22: IP Satellite Receivers – Control

5.4.2.3   Operating Platform

Historically, the operating system has been a limiting factor in the development of BTV/IP services. This was very apparent to anyone who attempted to use a printed circuit board style satellite receiver for personal computers such as those manufactured in the mid-to-late 1990s. Networks based on these technologies found the operating systems of the time (such as Windows 95) to be unreliable and the PCs regularly required rebooting or having their operating systems reinstalled.

Today, this has been overcome by newer, more reliable, operating systems. This section discusses the operating systems for IP satellite receivers as well as three other parameters: power, form factor, and memory. Definitions are included in Section 5.2.2.6

IP SATELLITE RECEIVERS > PLATFORM

	IDC SRA2000	IPRICOT SC+	IPRICOT S500+	MAINSTREAM DVB+	NOVRA S75	SKYSTREAM 1600
Operating System	Linux	Net BSD	Net BSD	Net ARM	None	Linux
Power	Universal AC Power, Autosensing	External Adapter 220 V or 110 V	External Adapter 220 V or 110 V	External Adapter 220 V or 110 V	External Adapter 220 V or 110 V	220 V or 110 V
Form Factor	Desktop (Rack opt.)	Desktop	Desktop	Desktop	Desktop	Desktop

Table #23: IP Satellite Receivers – Platform

5.4.2.4   TCP/IP Support

Any BTV/IP Satellite Receiver sits on a local area network and must function as a network appliance using the TCP/IP protocols. In addition, the receiver must have router functionality in order to distribute the content coming off the satellite onto the LAN. This section compares seven TCP/IP related categories relevant to these products.

All of the comparison categories in this section are of importance in selecting an IP satellite receiver. Definitions of the comparison categories are included in Section 5.2.2.7.

IP SATELLITE RECEIVERS > TCP/IP

	IDC SRA2000	IPRICOT SC+	IPRICOT S500+	MAINSTREAM DVB+	NOVRA S75	SKYSTREAM 1600
RJ-45 Ports	1	1	1	2	1	1
IP Streaming	Yes	Yes	Yes	Yes	No	No
TCP/IP Acceleration	No	Yes	Yes	Yes	No	No
Multicast	Yes	Yes	Yes	Yes	Yes	Yes
IGMP Grouping	Yes	Yes	Yes	Yes	Yes	Yes
SNMP	Yes	Yes	Yes	No	No	Yes
Throughput	20 - 30 MBs	8 MBs	44 MBs	40 MBs	60 MBs	10 MBs

Table #24: IP Satellite Receivers – TCP/IP

88

5.4.2.5 Security

Security issues in a BTV/IP environment are more complex than they are in a traditional BTV environment. Traditional BTV networks viewed encryption and conditional access at the DVB layer, protecting the entire satellite carrier from the outside, but not the individual programs once they are de-encrypted and de-scrambled by the satellite receiver.

Traditional BTV networks also knew that the receiver was part of the same technology platform as the DVB equipment at the network's head end. As a result, the conditional access system could be imbedded in the receiver negating the need for peripheral security items such as smart cards.

Definitions for the comparison categories are included in Section 5.2.2.8.

IP SATELLITE RECEIVERS > SECURITY

	IDC SRA2000	IPRICOT SC+	IPRICOT S500+	MAINSTREAM DVB+	NOVRA S75	SKYSTREAM 1600
IP Security	No	IPCrypt, IPSec	IPCrypt, IPSec	Proprietary	No	IP CAS
DVB Layer	Yes	Optional	Optional	Proprietary	No	No
CA DRM	No	Optional	Optional	No	No	No
Smart Card	No	Optional	Optional	No	No	No

Table #25: IP Satellite Receivers – Security

5.4.2.6 Receiver Specifications

The satellite receiver specifications need to be considered during the decision making process. Satellite receivers have earned a reputation for being highly reliable. Although this is true, not all satellite receivers are the same and this section will provide value when selecting a technology platform.

Comparison Category Definitions are included in Section 5.2.2.9.

IP SATELLITE RECEIVERS > RECEIVER SPECIFICATIONS

	IDC SRA2000	IPRICOT SC+	IPRICOT S500+	MAINSTREAM DVB+	NOVRA S75	SKYSTREAM 1600
RF Input	F-type, 75 ohm	F-type, 75 ohm	F-type, 75 ohm	F-type, 75 ohm	F-type, 75 ohm	F-type, 75 ohm
Input Level	-25 to 65 dBm	-25 to 65 dBm	-25 to 65 dBm	-25 to 65 dBm	-25 to 65 dBm	-25 to 65 dBm
Freq. Range	950 - 2150 mHz	950 - 2150 mHz	950 - 2150 mHz	950 - 2150 mHz	950 - 2150 mHz	950 - 2150 mHz
LNB Power	13 - 18 VDC	14 -18 VDC	14 -18 VDC	14 -18 VDC	13 -18 VDC	13 - 18 VDC
LNB Control	22kHz	22 kHz	22 kHz	22 kHz	22 kHz	22kHz
FEC	Virterbi 1/2, 2/3, 3/4, 5/6, 7/8 Reed-Solomon (204, 188)	Virterbi 1/2, 2/3, 3/4, 5/6, 7/8 (autosense) Reed-Solomon (204, 188)	Virterbi 1/2, 2/3, 3/4, 5/6, 7/8 (autosense) Reed-Solomon (204, 188)	Virterbi 1/2, 2/3, 3/4, 5/6, 7/8 (autosense) Reed-Solomon (204, 188)	Virterbi 1/2, 2/3, 3/4, 5/6, 7/8 (autosense) Reed-Solomon (204, 188)	Virterbi 1/2, 2/3, 3/4, 5/6, 7/8 Reed-Solomon (204, 188)
Demodulator	QPSK	QPSK	QPSK	QPSK	QPSK	QPSK
Symbol Rate	256Ksps - 45 Msps	1 - 45 Msps	1 - 45 Msps	2 - 45 Msps	1.5 - 45 Msps	1.08 - 45 Msps

Table #26: IP Satellite Receivers – Receiver Specifications

Appendix A - Evolution of Business Television

The roots of business television (BTV) go back to the late 1970s, with much credit given to the broadcast and cable industries for technologies and applications. Since then, BTV has developed into a solid industry, providing or enhancing corporate and employee communications and enterprise training requirements for organizations that need to reach multiple geographically-dispersed locations, economically, with timely information.

This section provides an overview of key phases, developments and suppliers in the BTV industry. It tells a story which should help identify how, and with whom, to proceed as you endeavor to deploy the next generation of BTV/IP technology.

From AT&T's T45 Terrestrial Service to Satellite Delivered Television

In the early days of television, program feeds from the country's major television networks (ABC, CBS, NBC and PBS) were distributed throughout the country via the T45 (45 Mbps) broadcast video lines of AT&T. The T45 lines extended to all cities with local television stations affiliated with the major networks.

As it is today, network programming was broadcast from the production and news studios of each of the networks, located in New York City. TV stations in each local market were affiliated with one of the networks, rebroadcasting its programming. Additional programming and commercials were inserted at the local level.

In addition to trafficking the primary program feeds via AT&T, the networks also used the T45 service to bring local news stories and live sporting/news events from venues around the country back to the studios in New York. News content was edited and approved for daily news feeds to the local affiliates. Live sporting events or breaking news stories would be turned-around live for broadcast on local, regional or national levels, depending on the importance of the content or local interests.

Through the 1950s, 60s and much of the 70s, AT&T provided the only means to distribute the program feeds, which required extremely high bandwidth for the rich media content. AT&T provided a web of point-to-point T45 lines, connecting cities from the east coast to the west, including Hawaii and Alaska.

Not only was T45 service expensive, it was also susceptible to technical glitches and human error. Some may remember the days when network programs would be interrupted with a different program. On many occasions, viewers would receive the wrong sporting event or programs that weren't scheduled. In most cases, the program feeds leaving New York would arrive at the right locations.

In 1965, ABC, CBS and NBC formed a Joint Network Task Force (JNTF). JNTF's objective was to provide input to the FCC (Federal Communications Commission) and to satellite fleet operators on how broadcasting could use communications satellites. Dissolved in 1973, the JNTF, nevertheless, provided significant early direction for the satellite industry.

The introduction of satellite-delivered television in the 70s elevated the video quality and reduced the costs of delivery of TV programming. Since 1975, communication satellites have played a significant role in the operations of the local broadcast stations, cable television and direct-to-home (DTH) businesses. A major league baseball game served as the first live satellite-delivered TV program in the U.S., transmitted by an uplink in Dallas-Fort Worth; HBO and Scientific Atlanta dished up the first worldwide satellite broadcast of Frazier and Ali's historic *Thrilla in Manila*, on October 1, 1975.

Before long, the three major networks began delivering sporting events and news stories to New York from satellite uplinks across the U.S.; and in the mid 80s, converted their primary TV program feeds from terrestrial delivery to satellite.

Eventually, a number of independent television stations throughout the country were purchased by, or merged with, new satellite-based networks such as FOX, Warner Brothers (WB) and United Paramount Network (UPN), all of which use satellite for the delivery of their primary programming services.

Broadcast & Cable Industry: Whence BTV Came?

The Public Broadcasting System (PBS) was the first nation-wide network to recognize the cost benefits of distributing it programming via satellite to local affiliates versus the traditional terrestrial lines. The three commercial networks (ABC, NBC and CBS) followed in the early 1980s, by going primarily to satellite for trafficking of backhaul feeds and program distribution.

Beginning in 1975, cable television, or "pay TV", came on the scene and became a major user of satellite delivery. As a perceived threat to "free" television, cable TV made significant inroads in rural areas, where most people were unable to receive local television signals. Cable TV was available via satellite for a monthly subscription fee, plus upfront hardware and installation costs. For the first time, satellite programming was received directly in the home.

Cable TV also became available in the metropolitan and more densely populated areas. Cable companies built out their network of local facilities to receive programming from the content providers. As one would expect, it would be quite expensive to traffic multiple channels of wide-band programming to numerous cities throughout the country, hence the need for cable companies to obtain the programming through less expensive means provided by satellite. From these facilities, the cable companies would downlink programming from various satellites and aggregate the distribution of the program signals over terrestrial cable lines. Backbone lines were laid underground to subdivisions and other residential areas, where separate lines were dropped into subscriber homes.

In parallel, satellite-based program subscription companies became interested in the opportunities presented by urban areas. A number of them attempted to launch Direct-to-Home (DTH) and Direct-Broadcast-Services (DBS). Two companies in particular survived the hotly contested marketplace: EchoStar's Dish Network service, which now enjoys delivering programming to its 9 million subscribers; and Hughes Electronics Corporation's DirecTV with the largest subscription base of 12 million.

92

BTV's Early Days... The Beginning!

In the early 1960s, AT&T previewed satellite-delivered video for its stockholders in a special closed circuit broadcast from Andover, Maine to Washington, D.C. The broadcast featured video of an American flag blowing, during a 20-minute window, as the AT&T Telstar satellite flew overhead on its 90-minute orbit of the earth.

Since that time, entrepreneurial companies have brought the benefits of the satellite-delivered education and training to the corporate environment, frequently using the same technologies developed for broadcasters and cable companies. Starting in the 1970s, ad-hoc Special Event Videoconferencing (SEV) developed into a viable communications tool for the enterprise market. Applications for satellite-based delivery of ad-hoc events grew rapidly for:

- Product introductions , such as automobiles and computers
- Time-sensitive statements to the media, such as Johnson & Johnson's 30-site event announcing the plan to recall Tylenol and its introduction of safety packaging
- Internal announcements, executive speeches, awards celebrations or other types of corporate events
- Customer and employee education and training, reducing the need for costly, time-consuming travel

At one time, there were dozens of companies servicing the ad-hoc Special Event market, including VideoStar Connections, Satellite Networking Associates, Satellease, Bonneville, Holiday Inn's Hi-Net Network, Wold Communications, VideoNet, and Netcom. Most were two-to-three person shops, working with what came to be known as third-party installers and field service personnel. VideoStar is now Convergent Media Systems (CMS). Bonneville and Wold are now known as GlobeCast North America. Vista Satellite Communications had its roots in SNS.

As today, many ad-hoc Special Events were viewed at hotels, theaters, convention centers and other public meeting facilities. They were typically serviced by rented transportable downlink systems and audio/visual (AV) equipment. The downlinks used C-band technology, which required large, cumbersome antennas. These C-band systems typically were mounted on boat trailers and became industry standard for the next 15-plus years. Using mobile equipment and a lot of creativity, companies were able to reach their audiences virtually anytime, anywhere.

Innovators included Marriott, Inc. and VideoStar, who teamed together to provide the Tele-Meeting Network. Marriott featured its meeting facilities, convention services and guest rooms and VideoStar handled the network and technical services. Holiday Inn established its HiNet Satellite Network, which provided an all-inclusive lower cost solution. Marriott continues to provide SEV services today, through its EventCom Technologies organization.

As companies grew comfortable with satellite-based services, they looked at how they could improve their organization's communications and/or training requirements by providing information on a regular, timely basis, by rolling transportable downlinks right to the end-user's remote office, retail or manufacturing facility, rather than to rented venues. Meeting at company facilities saved time and avoided the expense of room fees, A/V rentals and telephone installation and service. This activity marked the beginning of the business television industry.

Originally, the early providers of BTV broadcast over the lower frequency C-band and used its associated technologies. The C-band receive antennas were large, costly, and difficult to install, requiring complicated engineering specifications, structural modifications to building roofs, permits from the respective municipalities, and authorizations from building managers, landlords and property owners.

In the early 1980s, an alternative emerged: VSATs (Very Small Aperture Terminals). VSAT technology brought the satellite option to the transmission of data, enabling large-scale enterprises such as banks and government agencies to transmit and receive data by satellite. With the addition of a TV signal decoder, a two-way VSAT system could also receive one-way BTV programming. Many companies that installed VSAT networks to distribute data to multiple locations across the U.S. and Canada also used their networks for BTV.

VSAT systems used a different bandwidth, however, than the standard C-band TV transmission: Ku-band. The Ku-band technology, with its smaller dishes, quickly established itself as the standard for BTV, whether as an add-on to a VSAT network or as a stand-alone BTV network. Hewlett-Packard is recognized as having the first BTV network in 1983, and was soon followed by IBM and Tandem.

In 1986, Westcott Communications launched its Automotive Satellite Television Network (ASTN), marking the advent of vertical industry training networks. During this time, Westcott and other creative service providers introduced the non-penetrating roof mount (NPRM), which allowed for the installation of satellite dish mounts without penetrating the roofs of buildings. The NPRM simplified the installation process and made it economically feasible for more companies to install dedicated satellite networks throughout their organizations.

As vertical networks proliferated, NPRMs and Ku-band receive dishes sprung up on roofs across the U.S. As subscribers to ASTN, General Motors, Ford and Chrysler were early adopters of BTV, eventually using their ASTN networks to downlink their own programming. Other corporations followed suit, many subscribing to vertical training networks and then installing their own receivers to downlink private BTV broadcasts.

In the late 1980s, General Instruments and Scientific-Atlanta established themselves as leading providers of satellite receivers and decoders for the BTV industry, with their DigiCypher I and B-MAC technologies. Scientific-Atlanta quickly garnered a large share of the enterprise market as companies such as CMS and Telesat Canada's BTV Services Group chose the B-MAC for its customers (clients). Other networks, including the Ford Motor Company and those serviced by Canadian BTV service provider CanCom, selected General Instruments, later purchased by Motorola. It was during this period that the satellite receiver and decoder were integrated into a single unit known as the Integrated Receiver/Decoder (IRD).

Despite the success of business television, when compared to larger, more widely established industries such as telecommunications, broadcasting and cable, the satellite industry that supports BTV is still a cottage industry. Today, about 185 enterprise-wide BTV satellite networks operate in the U.S.

Business Television Goes Digital

AT&T Tridom led the digital push by introducing its Vistacast service in 1993, featuring the Compression Labs (CLI) SpectrumSaver digital system. AT&T Tridom embraced the digital system because of the cost savings it represented by allowing satellite transponder space segments to be subdivided into multiple channels. Many of the other service providers selected the SpectrumSaver technology for the same cost savings. SpectrumSaver was the only viable digital system available at the time.

General Instruments and Scientific-Atlanta came to market a couple of years later with their DigiCypher and PowerVu compressed digital systems. As with its B-MAC system, Scientific-Atlanta quickly garnered a large share of the enterprise market.

The 90s continued to be an excellent time for the development of video-based satellite communications, for both the enterprise and the consumer markets. During this period, DirecTV, EchoStar and AlphaStar came on line with their Direct Broadcast Services (DBS). They competed with cable television and the other lesser-established companies delivering entertainment programming to the home. All three companies entered the enterprise market to compete for BTV business, offering their low-cost, proprietary technologies and excess space segment.

EchoStar successfully acquired a share of the enterprise business, establishing itself as one of the key players in the industry. DirecTV and AlphaStar won a few enterprise customers. Soon thereafter, AlphaStar exited the scene due to a lack of subscribers.

In Canada, Shaw Communications introduced its StarChoice service. Shaw then purchased Canadian Communications (CanCom), merging it with StarChoice. In 2003, the BTV division was repurchased by the company's original founders and now provides services as BTV+.

By the mid-90s, the next generation of BTV technology appeared: IP multicasting. Demonstrated at trade shows, conferences and various locations, IP multicasting showcased its ability to economically and reliably deliver rich media content (video and multimedia applications and large data files) to multiple locations, something that terrestrial-based Internet or other WANs have yet to achieve.

In recent years, a number of satellite-based suppliers have provided their IP-based solutions to a handful of bleeding edge users. However, the market has been slow to embrace IP multicasting. For the most part, the enterprise market has elected to wait for applications solutions and providers to present a compelling story. The recession of 2001-02 has also taken a toll on the BTV industry, as corporations have pulled back on communication and training budgets and on capital spending.

On the other hand, it should be no surprise that the more aggressive, early users of media-rich IP systems are once again the broadcast networks. The major networks, and a selection of cable companies, have embraced satellite-based IP multicasting systems to deliver news, and in some cases, programming content, as well as advertising. In addition, there are a number of companies that have evolved from the Internet Service Provider (ISP) market, who compete in the enterprise space.

VSAT providers, as well, have adopted the use of IP-based solutions to support the transaction and data delivery requirements of many of their enterprise customers, and have migrated their BTV add-on services, where necessary, to support the one-way multicasting of media-rich content.

The long-standing business television providers, in many cases, have developed a clear understanding of the video and multimedia requirements of the enterprise customer, regardless of the application drivers: training; interactive distance learning; e-learning; merchandising, sales and marketing information; advertising, marketing and promotions (point-of-purchase, digital signage).

A distinct advantage for the satellite providers lies in their experience and longevity in the BTV market. Based on their experience, the traditional satellite-based service providers understand the video and multimedia capabilities driving business communications and training requirements of the typical enterprise communicator. What was broadcasting is now multicasting. What was live video is now streaming video. What was recording, is now caching. The technology and terminology has evolved; the services, fundamentally, remain the same.

The following table lists standard BTV terminology and its IP counterpart.

TV - PC CONVERGENCE TERMINOLOGY

TVs	PCs
Broadcasting	Multicasting
Live Video	Streaming Video
Networks	Portals
Channels	Sites
Producers/Providers	Content
Over-the-air	On-the-net
NTSC/PAL	IP/TV
Integrated Receiver/Decoder	IP Receiver
MPEG2	MPEG1 & 4
Remote Control	Interactive
VCR Recording	Caching/Storage/Catching/Fetching
Television	PC Monitor
Sound	Audio

Table #27 – TV-PC Convergence Terminology

End-user companies, organizations and associations continue to use satellite delivery of BTV programming on a full-time, as well as ad-hoc basis. The same applications that drove BTV in the early days of the industry are still key drivers today, including:

- **Time Management:** Maximizing executive and employee time and limiting the downtime associated with travel
- **Consistency of Message:** Delivering the same information simultaneously to a large, geographically-dispersed audience
- **Immediacy/Timeliness:** Distributing information in a timely manner
- **Increased Reach:** Getting the message to a widely dispersed audience
- **Cost Savings:** Limiting travel costs by broadcasting to audiences in numerous locations

Since the early days, BTV networks have matured. Users have taken advantage of technology advancements, market conditions, less expensive transmission and lower bandwidth space segment costs, and new applications to meet their communication requirements. Many now maintain full-time channels so that company information and news is available on a continuous basis, displayed in common areas, training and conference rooms, and delivered to the desktop.

The following table provides a "Then-Now" comparison between satellite-based "ad-hoc" events and webcasting.

BTV/AD HOC EVENTS — WEBCASTING	
THEN	**NOW**
Television Viewing	PC Monitor at the Desktop
Group Viewing in Conference Rooms	Video at the Desktop
Target Audience	Target Audience
• Employees	• Employees
• Clients	• Clients
• Suppliers	• Suppliers
• News Media	• News Media
• Wall Street/Financial Markets	• Wall Street/Financial Markets
Applications	Applications
• Training	• E-Learning
• Corporate Communications	• E-Communications
• Interactive Distance Learning	• Video-on-Demand (VOD)
• Meetings	• Web Meetings/Collaboration
• Sales & Marketing	• On-line Advertisements
	• E-Mail Advertisements

Table #28 – BTV/Ad Hoc Events - Webcasting

As the satellite footprints have expanded and the targeted workforce has changed, some companies are also using the satellite audio subchannels, as well as additional transmission channels, to transmit multilingual programming to Hispanic and French speaking viewers. Some things never change, however: companies continue to use Ad-hoc events and rented receive site equipment to reach extended and/or exclusive audiences.

DVB, MPEG-2, and the Birth of IP

In November 1994, the International Telecommunications Union (ITU) recommended that the DVB-S specification be adopted for Digital Satellite Television. This standard allowed manufacturers to develop products that were market-driven, interoperable, and flexible. When coupled with more powerful satellites requiring smaller satellite antennas, this standard rapidly resulted in the worldwide acceptance (as shown in the following diagram) and affordability.

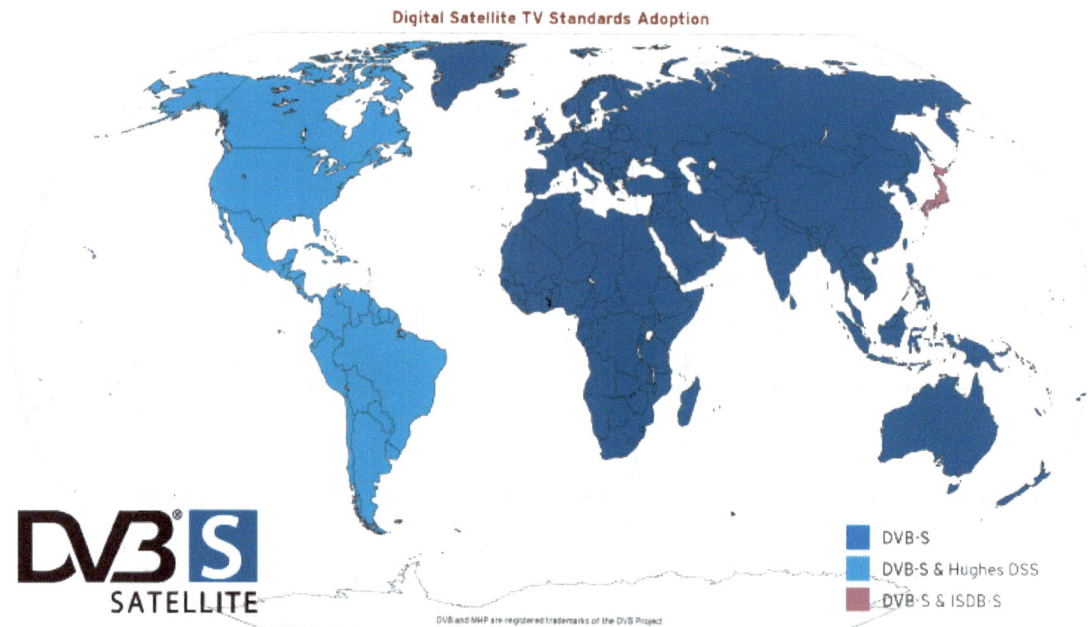

Source: The Digital Video Broadcasting Project (www.dvb.org)

Figure #19 - Digital Satellite TV Standards Adoption

DVB-S stands for "Digital Video Broadcasting via Satellite". This specification was developed by the European Launching Group (founded in 1991; later known as the DVB organization). The primary mission of this group was to develop a pan-European terrestrial digital television platform. However, due to a combination of technical and regulatory issues, satellite and cable delivery developed at a more rapid pace. This is particularly true of satellite as European satellite service providers were eager for an all-digital replacement for MAC technology.

One of the key elements of the DVB-S specification was the selection of MPEG-2 as the video encoding platform. The MPEG-2 standard (approved by the ISO in November 1993) provided broadcast quality digital video requiring a fraction of the bandwidth necessary for analog video transmission. In addition, MPEG-2 programming would be transported in an MCPC (Multiple Carriers Per Channel) configuration allowing for even greater bandwidth reduction on a carrier that could saturate a satellite transponder, thereby reducing the size of the receive antenna.

At the heart of the MPEG-2 standard was the MPEG-2 data packet containing the digitized video information for transport. These large data packets proved to be an excellent container for smaller IP data packets to travel over satellite. This capability allowed MPEG-2 to be the "Trojan Horse" for computer networking to merge with broadcast television.

Early MPEG-2/DVB Satellite Receivers for Business Television had opportunistic data ports, which allowed for some additional data services to be transported along with the video. Today, these opportunistic data ports have been upgraded to IP data ports.

In the late 1990s, companies such as Broadlogic and Gilat introduced Satellite Receivers integrated with PC cards. This allowed a Personal Computer to interface directly with a Satellite Antenna to provide low-resolution video and data to be downloaded at very high data rates. Services such as Hughes' DirectPC were introduced to satisfy this emerging market.

Problems encountered with these systems resulted from inferior computer technologies. In particular, Windows 95/98 was not an effective operating system for these applications and the Intel Pentium II was unable to process higher-resolution video. Unix based solutions were not an option as they were too costly for large scale implementations.

Since the late 90s, both satellite and computer technologies, and the ways in which they converge, have greatly improved. The Linux operating system has given rise to the IP Satellite Receiver Appliance, which today is networked on LANs to PCs with greater processing capabilities and more robust operating systems, allowing for the transport of higher impact video and multimedia content.

All of these developments are a result of the insight of the developers of DVB-S and MPEG-2.

Current Industry Overview

Like many other high-tech industries, the satellite industry has experienced its share of change and turnover during the recent recession. Despite the tough economic times, developments in the BTV space will likely have a positive impact on end-users since their last purchase or Request For Proposal (RFP).

Here's a brief satellite communications update:

- Launches of new satellites have continued, albeit at a reduced pace, increasing space segment capacity
- Significant advances in compression technology have resulted in increased space segment availability
- The size of satellite receive dishes continues to shrink, as the latest broadcast frequency, Ka-band, is poised to enter the market
- Satellite entertainment programming has exceeded 20 million subscribers through EchoStar and DirecTV, and is now recognized by many customers as providing better service than basic cable
- Mergers and acquisitions are extending the reach of BTV; examples include SES Global's purchase of GE Americom, merging its extensive resources into one of the largest, truly global companies; and Loral's sale of its North American satellite fleet and assets to Intelsat

In addition, satellite carriers have ventured beyond their traditional space segment services and now offer systems consulting, design and integration; many provide ground services as well. Intelsat has split its services into three business segments, including a Media & Entertainment Group that addresses broadcast and video applications. PanAmSat continues to address the video market, creating G2 Satellite Solutions from Hughes Global Services.

In a slightly different, but relatively close industry: SES Americom introduced its Americom2Home service, a broadband initiative in North America. Intelsat will compete in this market with its investment in WildBlue Communications.

The enterprise end-user benefits from this dynamic, robust competitive environment. Space segment is readily available and costs have come down. It's a buyer's market!

Business Television Service Providers - Market Sizing

Throughout BTV's history, the industry has experienced its share of mergers and acquisitions of service providers. Companies have been shut down or gone out of business. However, the industry has endured. In fact, the BTV industry has experienced very little recent change in service providers. Hence, a degree of stability:

- Convergent Media Systems maintains its position as an industry leader, after 23 years of service (Incorporated in 1980 as VideoStar Connection, Inc. and as EDS Video Services)
- GlobeCast North America does the same, after more than 30 years of service (back to the early 1970s of Wold Communications and Bonneville Communications, then as Keystone Communications, prior to being purchased by GlobeCast, a France Telecom company)
- Broadcast International has been providing BTV services for 20 years
- EchoStar's DISH Network Business Solutions remains a viable provider, since launching in the mid-1990s
- After 23 years in the industry, BTV+ is under the direction of the original founders/owners of Canadian Communications (CanCom), upon purchasing the company back from Shaw Communications
- Loral Skynet services more than 20 BTV clients through the merging of Loral's CyberStar into Skynet, including Global Access/Satellite Management Inc. and Satellite Network Systems (SNS) clients
- PanAmSat's G2 Satellite Solutions includes BTV clients from what was previously its sister company, Hughes Global Services. G2 supports government organizations and agencies

An estimated 185 corporate and government BTV networks operate today in North America. This is somewhat less than in the heyday of BTV, when there were well over 200 networks. This reduction is due in part to enterprise consolidations, acquisitions and mergers. Here is a breakdown of current network servicing in the U.S. and Canada:

- About 65 networks are serviced by Convergent Media Systems
- Loral Skynet (formerly known as Loral CyberStar) supports about 15 networks
- EchoStar manages about 40 networks, some of them through other service providers such as Convergent Media Systems and Broadcast International, as well as regionally-based systems integrators
- GlobeCast NA manages about 15 networks
- PanAmSat's G2 Satellite Solutions supports 7 networks, primarily government departments and organizations
- Broadcast International accounts for 6 separate networks, but supports a number of other clients in conjunction with Echostar
- BTV+ (formerly CanCom), in conjunction with Vista Satellite Communications, supports 7 networks in Canada

The remainder of the networks is spread over other service companies or self-managed. Educational, content delivery and vertical industry networks are not included in this sizing of business television networks.

The companies offer a selection of technologies and a wide range of services, packaged in different ways. This variety offers end-users a choice as they determine how best to meet their specific delivery and communication requirements. A Historical Perspective of BTV Service Providers is included in this guide.

A Chronology of BTV Service Providers

The following is a chronological list of companies providing business television and distance learning network services, offering an overview of the industry's evolution.

1970s   Wold Communications launched
Bonneville Satellite launched

1979   Foundation Telecommunications, Inc. founded
United Video launches Chicago International Teleport

1980   VideoStar Connections launched on April 1 to provide ad-hoc, occasional-use services
EchoStar launched to provide cable programming

1981   Canadian Communications (CanCom) launched

1983   Broadcast International, a music company, expands to include video services
Hughes Galaxy launched

1984   PanAmSat Corporation launched

1985   Miralite Communications launched
Chicago International Teleport renamed SpaceCom Systems

1987   GTE Spacenet launched
Hughes Network Systems pursues video business
VISTA Satellite Communications launched
Estimated launch of Satellite Management International
AT&T Skynet launched

1988   Microspace Communications launched

1989   VideoStar Connections bought by EDS; renamed EDS Video Services
Keystone Communications formed from merger of Wold Communications and Bonneville Satellite
Telesat Canada enters BTV space with Telesat Enterprises BTV Services Group

1992   AT&T Skynet folded into AT&T Tridom VistaCast
Group W Network Services launched
EDS Video Network Services goes private as Convergent Media Systems (CMS)

1993   Private Satellite Network (PSN) bought by CMS

1994   Digital Express launched
Muzak introduces its enterprise video network offering

1995   Broadcast International acquired by Data Broadcast Corporation (DBC)

1996   EchoStar launches DISH Network Business Solutions (DNBS)
        GTE Viznet launched
        Satellite Management Inc. bought by Vyvx, Global Access
        GlobeCast North America created by France Telecom's GlobeCast purchase of Keystone
        Communications

1997   AT&T Tridom VistaCast sold to GE Spacenet
        CyberStar launched by Loral
        Global Access bought by Loral CyberStar
        Hughes Global Services is launched by Hughes Network Systems
        Keystone Communications purchased by GlobeCast North America
        PanAmSat launched through merger of PanAmSat Corporation and Galaxy (of Hughes
        Communications)

1998   Network Group launched
        GTE Viznet shuts down BTV/IDL operations
        Group W ceases new sales development, continues to support BTV customers through the
        Liberty Media organization, Ascent Media
        GlobeCast North America purchases Hero Productions
        Digital Express ceases operations
        Muzak withdraws from video business

1999   Cancom merges with StarChoice, a Shaw Communications Company
        Miralite becomes incorporated as BitCentral
        Broadcast International sells audio services division (AEA) to Muzak; management team
        purchases BTV business from DBC
        Satcom Systems launched
        Telesat U.S. is launched

2000   Digistar Networks, Inc. incorporated by merged companies Network Group and Satcom
        Systems
        Global Telemann enters marketplace

2001   Diversified Media Group purchases Digistar Networks
        Global Telemann repositioned as Synoro
        AT&T purchases TCI, launches Digital Media Center

2002   Globecomm Systems Inc. enters marketplace
        Loral CyberStar merges into Loral Skynet
        Cancom separates from StarChoice and Shaw Communications and renamed BTV+
        Comcast purchases Digital Media Center as part of the cable acquisition from AT&T

2003   Hughes Global Services merges into PanAmSat as G2 Satellite Solutions
        MCI Satellite Services introduced (includes consolidation of IDB and other services)

Historical Perspective of Business Television Service Providers

COMPANIES	1979–2004 timeline
Ascent Media	Ascent Media
AT&T Tridom Vistacast	AT&T Skynet folded into Tridom/Vistacast — Purchased by GE Spacenet
BitCentral	Miralite Communications — Renamed as BitCentral
Broadcast International	Purchased by DBC — Split from Muzak
BTV+	Canadian Communications (CanCom) — StarChoice — BTV+
Comcast Digital Media Centers	AT&T DMC — Comcast
Convergent Media Systems	VideoStar Connections — EDS Video Networks — Convergent Media Systems
Digistar Networks, Inc.	Purchased by Diversified
Digital Express	
DirecTV	
Diversified Media Group	Networks Group — Digistar — Diversified Media
EchoStar DISH Network (DNBS)	
GlobeCast N.A.	Bonneville Communications — Keystone Communications — GlobeCast North America
Globecomm Systems Inc.	
Group W	See Ascent Media
GTE SpaceNet	Purchased by GE
GTE Viznet	
Hughes Global Services	PanAmSat
Hughes Network Systems	
Loral Skynet	Loral CyberStar — Loral Skynet
MCI	
MicroSpace Communications	
PanAmSat/G2 Satellite Solutions	PanAmSat Corporation — Merged with Galaxy (of Hughes Com.) — G2 Solutions
Primedia	Westcott — Primedia
Private Satellite Network	Purchased by CMS
Satellite Management Int'l	Purchased by Williams, Renamed Global Access — Purch. by CyberStar
Satellite Network Systems	Purch. by CyberStar
SpaceCom Systems	Chicago Int'l Teleport (under United Video, Inc.) — SpaceCom Systems — Pursued by EchoStar
Spacenet (Gilat Satellite)	GTE Spacenet — GE Spacenet — Spacenet (Gilat Satellite)
Telesat	Telesat Enterprises — Telesat US
Vista Satellite Communications	
Wold Communications	Merged with Bonneville to form Keystone Communications

TECHNOLOGY ADVANCES	
C-Band	
Analog Technologies	
Ku-band	
Compressed Digital Technologies - MPEG-1	
DVB Technologies - MPEG-2	
IP Technologies	
PC/Windows-based — Appliance/Linux-based	
Ka-band	

ANTENNA SIZING	
2.4+m	1.8 — 2.4+m	1.2 — 1.8+m	sub-meter — 1.8+m

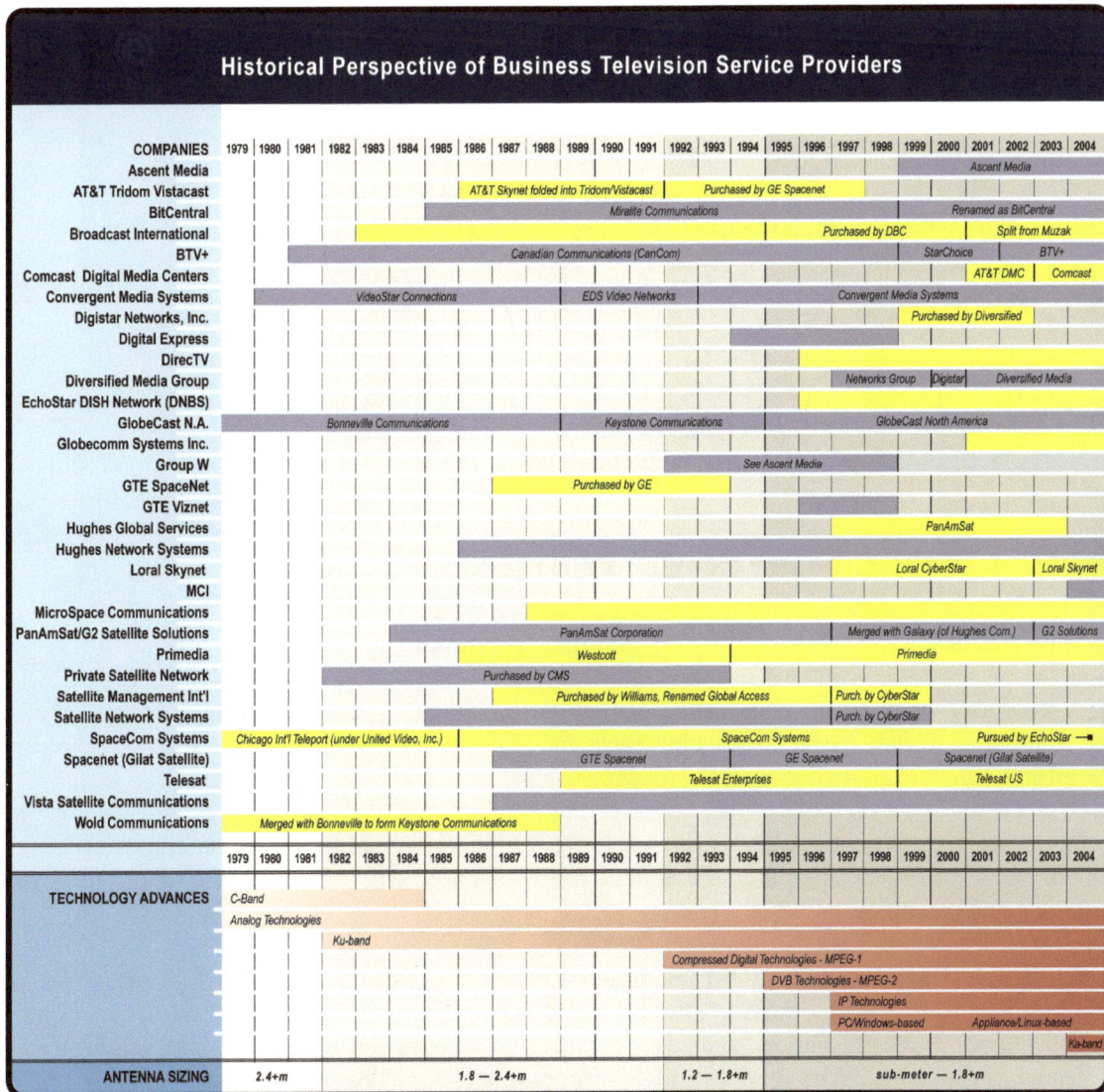

Figure #20 – Historical Perspective of Business Television Service Providers

Many VSAT service providers also support BTV networks as add-on enhancements to the two-way VSAT technology. Here's a chronological story of some of the key VSAT companies.

1986   Hughes kicks off VSAT industry with the Wal-Mart network
        Equitorial launched
        Comsat TP launched
        Contel launched
        Tridom launched
        AT&T/Harris launched
        ViaSat, Inc. launched

1987   AD/COM launched
        GTE Spacenet launched

1988   AD/COM purchased by Scientific-Atlanta

Tridom purchased by AT&T
AT&T/Harris merged with AT&T Tridom

1990   Equitorial purchased by Contel
        Comsat TP purchased by Contel

1991   Contel purchased by GTE Spacenet

1994   GTE Spacenet purchased by GE

1997   AT&T Tridom purchased by GE Spacenet

1999   GE Spacenet purchased by Gilat Satellite
        Scientific-Atlanta's VSAT Group purchased by ViaSat

Evolution of the Industry Service Provider

The BTV business and its service providers continue to evolve as a success story.

Major corporations over the years have pursued various interests in the satellite industry, whether they were video or data-based. Their objectives and expectations were that satellite could help drive new business opportunities and increase revenues.

In certain cases, this may be valid. However, some ventures have not provided the immediate results desired. As a result, the satellite interests have been dispensed with or folded into other businesses. Here are some examples:

Electronic Data Systems (EDS) purchased VideoStar Connections in 1989. At the time, VideoStar was a leading provider of business television networks (BTV) and services. As a subsidiary of General Motors, and sister-company to Hughes Network Systems (the industry leader in two-way VSAT services), EDS's intent was to supplement its Information Systems outsourcing services with satellite-based video communications and training capabilities.

VideoStar was re-named EDS Video Networks. By early 1993, EDS spun Video Networks off as Convergent Media Systems. Convergent Media operates today as a privately held company and maintains its position as a leading provider of BTV networks and related services.

AT&T purchased Tridom, one of the leading two-way VSAT companies, in 1989. During the mid-80s, terrestrial providers were losing data business to the satellite-based service providers, such as Tridom, HNS, GTE Spacenet and Scientific-Atlanta. VSATs provided a low-cost, easy to install alternative to AT&T and other terrestrial carriers. In 1992, AT&T moved the ground-based, video services group of its Skynet division to AT&T Tridom and re-named it Vistacast Services.

In 1997, when AT&T divested itself of all of its non-core businesses, such as NCR and Lucent, it also sold its Telstar satellite fleet to Loral and its Tridom organization to GE Spacenet.

GE American Communications (Americom), an early player in the manufacturing and management of communications satellites, purchased GTE Spacenet in 1994. In 1997, it purchased the Tridom organization from AT&T.

Early in 1999, GE sold the majority of its Spacenet interests to Gilat Satellite and in 2001, GE sold Americom to SES Global.

Shortly after purchasing GE Spacenet, Gilat elected to focus on its two-way VSAT business and its upstart Starband company, which provides high-speed Internet access to the consumer. In the end, Spacenet departed from the video-based, one-way BTV/IDL business.

Vyvx, the terrestrial broadband subsidiary of the Williams Company, purchased Satellite Management Inc., a leading business television company in the mid/late 90s. The BTV group was renamed Global Access. Vyvx and Williams combined to purchase a number of local and regional teleports over the next few years, essentially, establishing a transmission/delivery titan.

In 1997, the Global Access group was sold to Loral CyberStar.

In addition to Global Access, Loral CyberStar purchased Satellite Networks Services (SNS), another established business television company, primarily in the ad-hoc videoconferencing business. Both Global Access and SNS were folded into Loral CyberStar. In early 2001, all resources were consolidated with the company's headquarters in Rockville, MD. CyberStar elected to focus on its IP multicast business.

In 2002, CyberStar was merged into Loral's Skynet division, which continues to provide business television services.

Cable and broadcast titan, Group W, established a business television division in the early 1990s to generate additional revenue from its surplus resources: production studios, transmission facilities and space segment. By the late 90s, Group W elected to discontinue selling into the enterprise marketplace. Now, as part of the Liberty Media Group and Ascent Media, they continue to support BTV clients.

In 1999 Shaw Communications of Canada purchased Canadian Communications (CanCom), a business television service provider headquartered in Mississauga, Ontario, and merged it with its StarChoice organization, the Canadian direct-to-home satellite entertainment company.

In late 2002, the founders of CanCom purchased the BTV business back from Shaw, under the name of Canadian Business Television (CanBTV), more commonly referred to as BTV+.

In 2000, AT&T attempted to enter the business television arena again with the Digital Media Center (DMC) organization. It was based on using the video production and earth station resources in the Colorado facility, which was included in the purchase of the cable titan TCI. In 2002, AT&T exited the cable business by selling it to Comcast, including the DMC group.

Satellite-based video communications and multimedia delivery remains a vital industry. However, it's relatively small when compared to telecommunications and the Internet. The industry is not large enough to sustain and maintain the interest and ongoing support of major service providers.

History indicates that small to mid-size service providers may be better suited to nurture and thrive in the satellite-based industry. Of course, this may change if the enterprise market embraces IP multicasting over satellite delivery. The result will depend on how potential service providers from industries such as telecom, computer, software, video streaming, and visual display elect to become involved – whether through advertisers, training, distance education, corporate communications, or IT departments.

Service providers will continue to pursue the enterprise market, including satellite-based broadcast and cable industries; content management companies; router/server companies; streaming and ISP companies; training and e-learning companies; advertising, marketing and promotion companies and visual display (digital signage) companies.

What remains to be seen is which companies and solutions will acquire a foothold in the BTV/IP space.

Appendix B - Glossary of Terms

100BaseT- A 100-Mbps baseband Fast Ethernet specification using UTP wiring. Based on 10baseT technology and described in the IEEE 802.3 standard.

10BaseT- A 10-Mbps baseband Ethernet specification using two pairs of twisted pair cabling, (Category 3,4,or 5): one pair for transmitting data and one for receiving data. Part of the IEEE 802.3 specification, having a distance limit of 100 meters.

4:2:0- A sampling system used to digitize the luminance and color difference components (Y, R-Y, B-Y) of a video signal. The four represents the sampling frequency of Y, while the R-Y and B-Y are sampled between every other line only (one line is sampled at 4:0:0, luminance only, and the next at 4:2:2). This is generally used as a more economical system than 4:2:2 sampling for 625-line formats.

4:2:2- A commonly used term for a component digital video format. A ratio of sampling frequencies used to digitize the luminance and color difference components (Y, R-Y, B-Y) of a video signal. The term 4:2:2 describes that for every four samples of Y, there are two samples each of R-Y and B-Y, giving more chrominance bandwidth in relation to luminance compared to 4:2:0 sampling.

8PSK- Yields more throughput than QPSK by doubling the amount of data that can be encoded on the satellite carrier using the same amount of bandwidth.

16QAM- A more advanced type of Quadrature amplitude modulation. A complex but highly efficient downstream digital modulation technique that conforms to the International Telecommunications Union(ITU) standard ITU-T J. 83 which calls for quadrature amplitude modulation (QAM) with concatenated trellis coded modulation, plus enhancements such as variable interleaving depth for low latency in delay sensitive applications such as data and voice.

ARP (Address Resolution Protocol)- Defined in RFC 826, the protocol that traces IP addresses to MAC addresses.

Asynchronous transmission- Digital signal sent without precise timing, usually with different frequencies and phase relationships. Asynchronous transmissions generally enclose individual characters in control bits called start and stop bits that show the beginning and end of each character.

Back channel- In satellite communications, a terrestrial based communication link to the network hub, usually via internet or modem.

Backbone- The part of a network that acts as the primary path for traffic that is most often sourced from, and destined for, other networks.

BPSK- Biphase shift keying. BPSK is a digital frequency modulation technique used for sending data over a coaxial cable network. This type of modulation is less efficient--but also less susceptible to noise--than similar modulation techniques, such as QPSK and 16QAM.

BTV (Business Television) - Private corporate television networks used for internal communication of meetings, training and other business related content.

BTV/IP (Business Television over Internet Protocol) - Business Television content or programming delivered to the user over IP networks. BTV/IP networks have advantages over traditional BTV, such as storage of content, video on demand, delivery to the desktop and much more.

Category 5 cabling- One of five grades of UTP cabling described in the EIA/TIA-568B standard. Category 5 cabling is used for running CDDI and can transmit data at speeds up to 100 Mbps.

Composite video- An analog video signal that contains the audio and color subcarriers along with synchronization signals. In a television station this is how the signal appears just before being modulated (placed on a carrier) and sent to the transmitter.

DHCP (Dynamic Host Configuration Protocol) - A superset of the older BootP protocol, but includes certain enhancements. Both protocols use servers to dynamically configure clients when requested.

DMP - Dedicated Media Player

DNS (Domain Naming System) - System used in the Internet for translating names of network nodes into addresses. Sometimes can mean Domain Name Server.

Encapsulation- The technique used by layered protocols in which a layer adds header information to the Protocol Data Unit from the layer above.

EPG (Electronic Program Guide) - Provides an on-screen listing of all programming and content an interactive television service subscriber or digital television viewer has available to them. Usually manipulated by a hand held remote control device.

Ethernet- Baseband LAN specification invented by Xerox Corporation and developed jointly by Xerox, Intel, and Digital Equipment Corporation. Ethernet networks use CSMA/CD and run over a variety of cable types at 10 Mbps. Ethernet is similar to the IEEE 802.3 series of standards.

Fast Ethernet - Any of a number of 100-Mbps Ethernet specifications. Fast Ethernet offers a speed increase ten times that of the 10BASE-T Ethernet specification, while preserving such qualities as frame format, MAC mechanisms, and MTU. Such similarities allow the use of existing 10BASE-T applications and network management tools on Fast Ethernet networks. Based on an extension to the IEEE 802.3 specification.

FDDI (Fiber Distributed Data Interface) - LAN standard, defined by ANSI X3T9.5, specifying a 100-Mbps token-passing network using fiber-optic cable, with transmission distances of up to 2 km. FDDI uses a dual-ring architecture to provide redundancy.

Fiber-optic cable- Physical medium capable of conducting modulated light transmission. Compared with other transmission media, fiber-optic cable is more expensive, but is not susceptible to electromagnetic interference, and is capable of higher data rates. Sometimes called optical fiber.

FEC (Forward Error Correction) - A class of methods for controlling errors in a one-way communication system such as a satellite system. FEC sends extra information along with the data, which can be used by the receiver to check and correct the data. FEC can be one of many complex algorithms; the most popular are Reed-Solomon and Virterbi. The FEC ratio can be adjusted to compensate for noise on the satellite link.

Firewall- Router or access server, or several routers or access servers, designated as a buffer between any connected public networks and a private network. A firewall router uses access lists and other methods to ensure the security of the private network.

Firmware- Software instructions set permanently or semi permanently in ROM.

FTP (File Transfer Protocol) - The TCP/IP protocol used for transmitting files between network nodes such as severs and clients, it supports a broad range of file types and is defined in RFC 959.

GUI (Graphical User Interface) - User environment that uses pictorial as well as textual representations of the input and output of applications and the hierarchical or other data structure in which information is stored. Conventions such as buttons, icons, and windows are typical, and many actions are performed using a pointing device (such as a mouse). Microsoft Windows and the Apple Macintosh are prominent examples of platforms utilizing a GUI.

HTML (Hyper Text Markup Language) - A simple document formatting language that uses tags to indicate how a given part of a document should be interpreted by a viewer application such as a web browser.

HTTP (Hyper Text Transfer Protocol) - The protocol used by Web browsers and Web servers to transfer files, such as text and graphics file.

IGMP (Internet Group Management Protocol) - Used by IP hosts to report their multicast group memberships to an adjacent multicast router.

IP Address- 32-bit address assigned to hosts using TCP/IP. An IP address belongs to one of five classes (A, B, C, D, or E) and is written as 4 octets separated with periods (dotted decimal format). Each address consists of a network number, an optional subnetwork number, and a host number. The network and subnetwork numbers together are used for routing, while the host number is used to address an individual host within the network or subnetwork.

IRD (Integrated Receiver Decoder) - A device that receives a video signal satellite, demodulates and decodes the signal. The video information is then formatted to the correct standards to accommodate the attached video or network devices.

LAN (Local-Area Network) - High-speed, low-error data network covering a relatively small geographic area (up to a few thousand meters). LANs connect workstations, peripherals, terminals, and other devices in a single building or other geographically limited area.

Linux- Linux is a UNIX like operating system designed by Linus Torvalds. It is an open source (nonproprietary) product and thus is available with its source code. Linux is recognized as one of the most reliable and stable operating systems available today.

LMS (Learning Management System) - An application that tracks student activity and performance as they progress through an on-line or computer based curriculum.

LNA (Low Noise Amplifier) - An electronics package located at the satellite antenna that amplifies the weak signal levels from the satellite with no appreciable amplification of background noise.

LNB (Low Noise Block) - Generally a package that contains both LNA and LNC electronics. Note that in conversation, the terms LNA, LNB and LNC are often used interchangeably, although this is incorrect.

LNC (Low Noise Converter) An electronics package that converts the microwave frequencies of the satellite link to an intermediate frequency, usually between 950 – 1250 MHz

MAC Address- Standardized data link layer address that is required for every port or device that connects to a LAN. Other devices in the network use these addresses to locate specific ports in the network and to create and update routing tables and data structures.

Metadata- Data that describes other data. Metadata describes how and when and by whom a particular set of data was collected, and how the data is formatted. Metadata is essential for understanding and locating information stored in data warehouses and has become increasingly important in Web applications.

Network Address- Network layer address referring to a logical, rather than a physical, network device. Also called a *protocol address.*

NIC (Network Interface Card) Board that provides network communication capabilities to and from a computer system. Also called an *adapter.*

Packet- Logical grouping of information that includes a header containing control information and (usually) user data. Packets are most often used to refer to network layer units of data.

MPEG (Moving Pictures Experts Group) - A processing-intensive standard for data compression for motion pictures that tracks movement from one frame to the next and only stores the data that has changed.

MPEG v1- A group of picture blocks, usually four, which are analyzed during MPEG coding to give an estimate of the movement between frames. This generates the motion vectors that are then used to place the macroblocks in decoded pictures. This was designed to work at 1.2 Mbps, the data rate of CD-ROM, so that video could be played from CDs. However the quality is not sufficient for TV broadcast.

MPEG v2- Designed to cover a wide range of requirements from "VHS quality" all the way to HDTV through a series of algorithm "profiles" and image resolution "levels." With data rates of between 1.2 and 15 Mbps, there is intense interest in the use of MPEG-2 for the digital transmission of television--including HDTV--applications for which the system was conceived.

MPEG v4- The third standard developed by MPEG. Started in July 1993, provides a harmonized range of responses to the diverse needs of the digital audio-visual industry, including compatibility with other major standards such as H.263 and VRML.

NAT (Network Address Translation) - An algorithm used to minimize the requirements for globally unique IP address, permitting an organization using private (not globally unique) addresses to connect to the Internet, by translating those addresses into globally routable address space.

NIC (Network Interface Card) - A circuit board in a host such as a PC or server that provides access to a LAN. Also referred to as a Network Adapter.

NMS (Network Management System) - A wide variety of software applications and hardware products that help network system administrators manage a network. Network management covers a wide area, including: *Security*: Ensuring that the network is protected from unauthorized users. *Performance*: Eliminating bottlenecks in the network. *Reliability*: Making sure the network is available to users and responding to hardware and software malfunctions.

NTSC (National Television System Committee) - The organization that developed the analog television standard currently in use in the U.S., Canada, and Japan. Now it is generally used to refer to that standard. The NTSC standard combines blue, red, and green signals modulated as an AM signal with an FM signal for audio.

PAL (Phase Alternate Line) - The television broadcast standard throughout Europe (except in France and Eastern Europe, where SECAM is the standard). This standard broadcasts 625 lines of resolution, nearly 20 percent more than the U.S. standard, NTSC, of 525.

PAL-M - A derivative of the PAL standard used in most of South America.

PAL-N - A derivative of the PAL standard used in part of South America including Argentina, Paraguay and Uruguay.

PID – (Program Identifier) - Used to select program content from a multiplexed stream of many programs.

PVR   Personal Video Recorder

QPSK (Quadrature Phase Shift Keying) - QPSK is a digital frequency modulation technique used for sending data over coaxial cable networks. Since it's both easy to implement and fairly resistant to noise, QPSK is used primarily for sending data from the cable subscriber upstream to the Internet.

Rack Unit- An industry standard used to describe how much space a device occupies in a standard 19-inch wide equipment rack. A rack unit is 1.75 inch high.

RJ Connector   (Registered jack connector) - Standard connectors originally used to connect telephone lines. RJ connectors are now used for telephone connections and for 10BASE-T and other types of network connections. **RJ-11, RJ-12,** and **RJ-45** are popular types of RJ connectors.

Router- Network layer device that uses one or more metrics to determine the optimal path along which network traffic should be forwarded. Routers forward packets from one network to another based on network layer information. Occasionally called a *gateway* (although this definition of gateway is becoming obsolete.)

RS-232 - Common physical layer interface standard, developed by EIA and TIA that supports unbalanced circuits at signal speeds of up to 64 kbps. Closely resembles the V.24 specification. An example would be a serial port on a PC. Also known as **EIA/TIA-232**

SDI (Serial Digital Interface) - The standard based on a 270 Mbps transfer rate. This is a 10-bit, scrambled, polarity independent interface, with common scrambling for both component ITU-R 601 and composite digital video and four channels of (embedded) digital audio. Most new broadcast digital equipment includes SDI, which greatly simplifies installation and signal distribution. SDI uses standard 75-ohm coax cable and BNC connector. SDI can transmit the signal over 600 feet (200 meters) depending on cable type.

Server- Node or software program that provides services to clients.

SNMP (Simple Network Management Protocol) - This protocol polls SNMP agents or devices for statistical and environmental data. SNMP works with MIB that are present in the SNMP agent. This information is queried and sent to the SNMP server.

Streaming Media- Multimedia content--such as video, audio, text, or animation--that is displayed by a client a client, such as a PC, as it is received from the Internet, broadcast network, or local storage.

S-Video (Super-Video)- A technology for transmitting video signals over a cable by dividing the video information into two separate signals: one for color (chrominance), and the other for brightness (luminance). When sent to a television, this produces sharper images than composite video. (The terms **Y/C video** and **S-Video** are the same.)

Synchronous Transmission- Digital signals that are transmitted with precise clocking, such that clocking circuits at the receiver match the exact frequency of the controlling clock at the transmitter. This mode of transmission requires no start and stop bits.

TCP/IP (Transmission Control Protocol/Internet Protocol) - Common name for the suite of protocols developed by the U.S. DoD in the 1970s to support the construction of worldwide internet works

TFTP (Trivial File Transfer Protocol) - A stripped down version of FTP, it has no directory or browsing features. It can do nothing but send and receive files.

Transparent Proxy - Provides a configurable caching server that is completely invisible to all users of your LAN or ISP. The cache provides a substantial savings of network bandwidth.

UDP (User Datagram Protocol) - A connectionless transport layer protocol that allows datagrams to be exchanged without acknowledgements or delivery guarantees, relying on other protocols to handle error processing and retransmission. Described in RFC 768.

UNIX - Operating system developed in 1969 at Bell Laboratories. UNIX has gone through several iterations since its inception. These include UNIX 4.3 BSD (Berkeley Standard Distribution), developed at the University of California at Berkeley, and UNIX System V, Release 4.0, developed by AT&T.

USB (Universal Serial Bus) - A bus that is expected to replace serial and parallel ports, designed to make installation and configuration of I/O devices easy, providing for as many as 127 devices on a host. The USB architecture requires only one set of system resources for all devices attached to the bus.

VoIP (Voice over Internet Protocol) - Digitized voice encapsulated in IP packets and transported across a LAN or WAN.

VPN (Virtual Private Network) - Uses IP tunneling and encryption to provide security across a shared network such as an ISP. To the end user the network appears as a private LAN.

WAN (Wide-Area Network) - Data communications network that serves users across a broad geographic area and often uses transmission devices provided by common carriers. Frame Relay, ATM, and X.25 are examples of WANs.

Web Browser Interface- GUI-based hypertext client application, such as Mosaic, Internet Explorer or Netscape, used to access hypertext documents and other services located on innumerable remote servers throughout the WWW and Internet.

WINsock (Windows Socket Interface) - A software interface that makes it possible for an assortment of applications to use and share an Internet connection.

Enliten Management Group, Inc.
Atlanta, GA

enliten.net

www.ingramcontent.com/pod-product-compliance
Lightning Source LLC
Chambersburg PA
CBHW041446210326
41599CB00004B/154